Multiferroic Materials

Dr. R. Saravanan, M.Sc., M.Phil., Ph.D.

Associate Professor
Research Centre and PG Department of Physics
The Madura College (Autonomous)
Madurai – 625 011

Published as part of the book series
Materials Research Foundations
Volume 140 (2023)
ISSN 2471-8890 (Print)
ISSN 2471-8904 (Online)

Print ISBN 978-1-64490-226-4
ePDF ISBN 978-1-64490-227-1

This book contains information obtained from authentic and highly regarded sources. Reasonable efforts have been made to publish reliable data and information, but the authors and publisher cannot assume responsibility for the validity of all materials or the consequences of their use. The authors and publishers have attempted to trace the copyright holders of all material reproduced in this publication and apologize to copyright holders if permission to publish in this form has not been obtained. If any copyright material has not been acknowledged, please write and let us know so we may rectify in any future reprint.

Distributed worldwide by

Materials Research Forum LLC
105 Springdale Lane
Millersville, PA 17551
USA
http://www.mrforum.com

Printed in the United States of America
10 9 8 7 6 5 4 3 2 1

Table of Contents

Preface

Multiferroic materials, in which two or more ferroic ordering take place in the same phase have been studied extensively on fundamental and technological aspects. The possibility of controlling electrical polarization by a magnetic field and magnetization by an electric field (known as magnetoelectric effect) in these materials can give potential applications which make use of coupling between magnetic and electric orders. Multiferroic materials are technologically more capable due to their potential applications in data storage, spin valves, spintronics, memories, sensors and microelectronic devices. The multiferroic materials that exhibit ferromagnetism and ferroelectricity in a single phase are rarely available. Among all single-phase perovskite-type (ABO_3) multiferroic materials such as $BiMnO_3$, $LaMnO_3$ and $BiFeO_3$, etc., lanthanum orthoferrite ($LaFeO_3$) is an extensively investigated material worldwide. However, in $LaFeO_3$, the coupling of ferroic orderings is usually too weak to be practically applicable. By integrating various cations in the La/Fe site of $LaFeO_3$, coupling of magneto-ferroelectric orderings and hence the physical properties can be greatly modified. In this work, divalent, trivalent and tetravalent cations have been substituted in the place of La^{3+} ion and the modified physical properties have been analyzed. The pure and doped $LaFeO_3$ have important properties like high electrical conductivity, outstanding thermal stability, high dielectric constant, low dielectric loss, ferromagnetism and ferroelectricity. Because of these properties, these materials have been used in solid oxide fuel cell (SOFC), magneto-hydrodynamic power generation (MHD), capacitors and energy storage devices in microelectronics, non-volatile magnetic memory devices and ferroelectric random access memories (Fe-RAM).

The study of precise geometrical structure, interatomic chemical bonding and charge (electron) density distribution provides knowledge on the physical and chemical properties of a material which leads to the development of new devices and applications. However, the detailed analysis of interatomic chemical bonding and charge density distribution is lacking in literature for many important materials including lanthanum orthoferrite multiferroics. The present research deals with the charge density distribution studies of four series of lanthanum orthoferrite (LFO)-type multiferroic materials using experimental X-ray diffraction data. An investigation of the results obtained from other characterization works like scanning electron microscopy (SEM), energy dispersive X-ray spectroscopy (EDS), UV-visible spectrophotometry (UV-Vis), vibrating sample magnetometry (VSM), dielectric and ferroelectric measurements has also been carried out in this work.

Chapter I gives the objectives of the present work. It gives a detailed literature review on multiferroic materials, structure of $LaFeO_3$, properties and applications of pure and substituted LaFeO$_3$. It explains the solid state reaction method and the synthesis procedure to synthesis four series of lanthanum orthoferrite-type multiferroics required for this work. It explains the working principles of the various characterization techniques such as powder X-ray diffraction, scanning electron microscopy (SEM), energy dispersive X-ray spectroscopy (EDS), UV-visible spectrophotometry (UV-Vis), vibrating sample magnetometry (VSM), dielectric and ferroelectric measurements. This chapter explains the fundamental information about powder X-ray diffraction (PXRD) profile fitting technique, and also describes the methodology for estimating the experimental charge density. The maximum entropy method which is the most precise model to estimate charge density distribution is explained in this chapter in a detailed manner with related methodology. It also describes the methodologies for estimating the energy band gap and grain size of the samples.

Chapter II presents the results obtained from various characterization techniques and analytical techniques performed to investigate the four series of lanthanum orthoferrite-type multiferroics. The plots of experimental X-ray diffraction patterns, Rietveld fitted profiles, SEM micrographs, EDS spectra, UV-visible absorption spectra, Tauc plots, magnetization versus magnetic field (M-H) curves, dielectric plots and ferroelectric loops, 3-dimensional (3D), 2-dimensional (2D) electron density contour maps and 1-dimensional (1D) electron density line profiles are presented in this chapter. The Tables of structural parameters refined from the unit cell refinement method, Rietveld method and the elemental compositions of the prepared multiferroics from EDS analysis are also presented. The Tables of Bond angles and bond lengths in the host lattice, bond lengths and electron density values at the bond critical point from MEM analysis, energy band gap values, parameters of magnetic, dielectric and ferroelectric measurements are also presented.

Chapter III deals with the detailed analysis of results obtained from various characterization techniques for all the synthesized multiferroics. The interpretation of results, comparison of physical properties and charge density distribution for all the synthesized samples are also discussed elaborately. The optical, magnetic, dielectric and ferroelectric properties of the multiferroics have also been investigated through charge density values.

Chapter IV presents the major conclusion of the findings of the reported work.

Articles published based on the results in International Journals

[1] Interatomic chemical bonding and charge correlation of optical, magnetic and dielectric properties of $La_{1-x}Sr_xFeO_3$ multiferroics synthesized by solid-state reaction method, *G. Gowri,* R. Saravanan, S. Sasikumar, M. Nandhakumar, R. Ragasudha, *Journal of Material Science: Materials in Electronics*, (Springer Publication), Vol.30, 4409-4426, (2019) (I.F: 2.22).

[2] Exchange bias effect, ferroelectric property, primary bonding and charge density analysis of $La_{1-x}Ce_xFeO_3$ multiferroics, *G. Gowri,* R. Saravanan, S. Sasikumar, I.B. ShameemBanu, *Materials Research Bulletin*, (Elsevier publication), Vol.118, 110512, (2019), (I.F: 4.019).

[3] Investigation on interatomic chemical bonding and charge-related optical, multiferroic properties of $La_{1-x}Zn_xFeO_3$ bulk ceramics, *G. Gowri,* R. Saravanan, N. Srinivasan, O.V. Saravanan, S. Sonai, *Materials Chemistry and Physics*, (Elsevier publication), Vol.267, 124652, (2021), (I.F: 3.408).

[4] Probing the effects of Al dopant over the structure and charge related optical, magnetic and electrical properties of Al^{3+} doped $LaFeO_3$ bulk multiferroic materials, *G. Gowri,* R. Saravanan, N. Srinivasan, K. Karunya, P. Jeyasheela, M. Uthra, *Chemicalpapers*, (Springer Publication), Vol.75, 4337-4353 (2021), (I.F: 1.68).

[5] Interatomic bonding and charge ordering in superparamagnetic $La_{0.8}Zn_{0.2}FeO_3$ Multiferroic, *G. Gowri,* R. Saravanan, *International Journal of Latest Trends in Engineering and Technology,* Special Issue - International Conference on Nanotechnology: The Fruition of Science-2017, 144-150, (2017), e-ISSN:2278-621X. DOI: 10.21172.

List of papers presented in conferences

[1] Synthesis and electron density analysis of $LaFeO_3$ multiferroic, *G. Gowri,* R. Saravanan, UGC sponsored one day International conference on "Recent Trends in Materials Science and Applications" organized by the Department of Physics, Sri Meenakshi Government Arts College for Women, Madurai-2, on January 6th 2016, ISBN : 819331402-6.

[2] Interatomic bonding and charge ordering in superparamagnetic $La_{0.8}Zn_{0.2}FeO_3$ multiferroic, *G. Gowri,* R. Saravanan, SERB-DST & MINISTRY of Earth Sciences, Government of India sponsored International Conference on "Nanotechnology: The Fruition of Science ICON-17" organized by Research committee, Nesamony Memorial Christian College, Marthandam, on 15th & 16th February 2017.

[3] Charge density and crystal structure analysis of $La_{0.85}Ce_{0.15}FeO_3$ ceramic, *G. Gowri,* R. Saravanan, N. Srinivasan, O.V. Saravanan, S. Sonai, CSIR sponsored National conference on "New-Generation Materials for Energy andApplications" conducted by the Department of Physics, B.S.Abdur Rahman Crescent Institute of Science and Technology, Chennai, on 21st & 22nd October 2019, ISBN 13: 978-81-942285-1-6.

Multiferroic Materials   Materials Research Forum LLC
Materials Research Foundations **140** (2023)   https://doi.org/10.21741/9781644902271

Chapter 1

Introduction

Abstract

In this work, the following four series of lanthanum orthoferrite (LFO)-type multiferroics have been synthesized by solid state reaction method.

(i) $La_{1-x}Ce_xFeO_3$ (x=0.00, 0.03, 0.06, 0.09 and 0.12) (LCFO)

(ii) $La_{1-x}Zn_xFeO_3$ (x=0.00, 0.05, 0.15, and 0.25) (LZFO)

(iii) $La_{1-x}Al_xFeO_3$ (x=0.05, 0.15, 0.25 and 0.35) (LAFO) and

(iv) $La_{1-x}Sr_xFeO_3$ (x=0.05, 0.10, 0.15 and 0.20) (LSFO)

The synthesized samples have been characterized using powder X-ray diffraction (PXRD), scanning electron microscopy (SEM), energy dispersive X-ray spectroscopy (EDS), UV-visible absorption spectroscopy (UV-Vis), vibrating sample magnetometry (VSM), dielectric measurements (Impedance analysis), ferroelectric measurements (polarization versus electric field (P-E) hysteresis loop). For all the synthesized multiferroics, the charge density distribution and bonding nature of atoms in the unit cell have been analyzed accurately using Maximum Entropy Method (MEM). The optical, magnetic, dielectric and ferroelectric properties have been analyzed using MEM-based BCP charge density values. The results obtained from the above experimental characterizations and analytical techniques are analyzed and summarized in this chapter.

Keywords

Multiferroic, Charge Density, X-Ray, SEM, EDAX, Optical, Electrical

1.1 Objectives

The objective of the present work is to synthesize and characterize four series of lanthanum orthoferrite (LFO)-type multiferroics with various dopant concentrations. These materials could exhibit (anti)ferromagnetism and ferroelectric properties at room temperature which would make them good candidates for practical applications. The objective also includes investigation on charge density distribution and the effect of substitution of various cations on the structural, optical, magnetic, dielectric and ferroelectric properties through charge density distribution which has not yet been carried out. In order to fulfill the objectives, the following tasks have been carried out.

1. Synthesis of lanthanum orthoferrite (LFO)-type multiferroics viz.,

   (i) $La_{1-x}Ce_xFeO_3$ (x=0.00, 0.03, 0.06, 0.09 and 0.12) (LCFO)

   (ii) $La_{1-x}Zn_xFeO_3$(x=0.00, 0.05, 0.15 and 0.25) (LZFO)

   (iii) $La_{1-x}Al_xFeO_3$(x=0.05, 0.15 and 0.25) (LAFO) and (iv) $La_{1-x}Sr_xFeO_3$ (x= 0.05, 0.10, 0.15 and 0.20) (LSFO)

2. Characterization of the synthesized lanthanum orthoferrite-type multiferroics using powder X-ray diffraction (PXRD) and structural analysis using the Rietveld [Rietveld, 1969] refinement technique by adapting the JANA 2006 [Petricek et al.,2014] software program, to examine the structural properties. Estimation of grain size using Scherrer [Cullity and Stock, 2001] formula using the GRAIN software program [Saravanan, 2008].

3. Analysis of the microstructure and surface morphology through the images recorded using a scanning electron microscope (SEM).

4. Identification of the elemental composition of the synthesized multiferroics qualitatively and quantitatively by Energy dispersive X-ray spectroscopy (EDS).

5. Study of the effects of substitution of various elements (Ce, Zn, Al and Sr) at the La site in the host system $LaFeO_3$ on the interatomic chemical bonding and charge density distribution through Maximum Entropy Method (MEM) [Collins, 1982] by adapting the software programs PRIMA [Izumi, 2002] and VESTA [Momma and Izumi, 2008].

6. Estimation of energy band gap (E_g) values of the synthesized multiferroics using UV-visible (UV-Vis) absorption spectra and Tauc plot (Wood and Tauc, 1972).

7. Investigation of magnetic properties of the multiferroics using vibrating sample magnetometer (VSM) measurements.

8. Study of variation of dielectric constant (ε'), dielectric loss factor (tan δ) and ac conductivity (σ_{ac}) as a function of frequency at room temperature for all the multiferroics using (Impedance analysis) dielectric measurements.

9. Study of ferroelectric properties of the synthesized multiferroics using polarization versus electric field (P-E) hysteresis loop measurements.

10. Analysis of optical properties and multiferroic properties such as magnetic and ferroelectric properties through charge density distribution has been done for the first time for the synthesized multiferroics.

The important features of $LaFeO_3$, the description of the above-mentioned experimental and analytical techniques, and the details of the synthesis of lanthanum orthoferrite-type multiferroics are presented in this chapter. The results obtained from various characterizations and analytical techniques, and the major conclusion drawn from the results are provided in the following chapters.

1.2 Multiferroic materials –An overview

1.2.1 Multiferroics and magnetoelectrics

Materials that exhibit two or more ferroic properties such as (anti)ferromagnetism/ferrimagnetism, ferroelectricity, ferroelasticity and ferrotoroidicity in a single phase are described as multiferroics [Spaldin et al., 2008, Fiebig et al., 2016]. They have the potential for applications as actuators, switches, magnetic field sensors and new types of electronic memory devices [Dong et al., 2015]. Multiferroic materials that exhibit both magnetic and electric ordering in a single phase are called magnetoelectrics. They have attracted great interest because of their technological applications for electronic devices and fascinating fundamental physics. The combination of ferromagnetism and ferroelectricity into a multiferroic is sketched in Figure 1.2(a) [Khomskii, 2009]. These magnetoelectric materials possess spontaneous magnetic polarization that can be reversed by a magnetic field and spontaneous electric polarizationthat can be switched by an applied electric field [Cheong and Mostovoy, 2007]. The combination of ferroelectricity and ferromagnetism arises from different mechanisms. The ferroelectricity arises from the electric dipole ordering, and the ferromagnetism arisesdue to the spin ordering. The origin of magnetism in multiferroics is due to the presence of localized electrons, mainly in the partially filled d or f shells of transition metal or rare-earth ions, which have a corresponding localized spin, or magnetic moment. Exchange interactions between the localized moments lead to magnetic ordering in multiferroic materials. The situation is quite different with ferroelectrics. FE polarization in a material originates from the distorted asymmetric crystal structure, but a symmetric structure is needed for ferromagnetism. The common mechanism for ferroelectricity is off-centering of B-site d^0 electrons. Therefore, for ferroelectricity and magnetism to occur simultaneously in a single phase, the atoms that move from the center to form the electric dipole moment should be different from those that carry the magnetic moment.

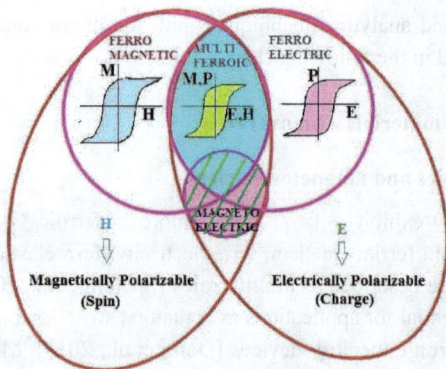

Figure 1.2(a) Multiferroics combine the properties of magnets and ferroelectrics

1.2.2 Magnetoelectric (ME) effect

The magnetoelectric effect denotes the coupling between electric and magnetic ordering in a material. It is characterized by the variation of an induced magnetization (M) by an applied electric field (E) or the electrical polarization (P) in response to an applied magnetic field (H) [Smolenskii, 1978] and is expressed in the following way.

$$P = \chi^E E + \chi_{ij}^{EM} H \qquad\qquad \text{......... (1.2.1)}$$

$$M = \chi^M H + \chi^{ME} E_{ji} \qquad\qquad \text{......... (1.2.2)}$$

where χ^E is electric susceptibility tensor, χ^M is magnetic susceptibility tensor,
$\chi^{EM} = \chi^{ME} = \delta^2\phi/\delta E\, \delta H$ is mixed magnetoelectric susceptibility tensor.

ij   ji   i   j

$$\chi^{EM} = \delta P/\delta H \qquad\qquad \text{......... (1.2.3)}$$

$$\chi^{ME} = \delta M/\delta E \qquad\qquad \text{......... (1.2.4)}$$

ϕ – thermodynamic potential.

1.2.3 Multiferroism and symmetry

As crystal symmetry plays a vital role in ferroelectric properties, it is clear that each multiferroic property is closely linked to symmetry. The primary ferroic properties can be characterized by their behavior under space and time inversion, as shown in Table 1.1. For example, the operation of space inversion will reverse the direction of polarisation (p), while leaving the magnetization (M) invariant. Therefore, non-polar ferromagnets and ferroelastics are invariant under space inversion, whereas polar ferroelectrics and ferrotoroids are not. The operation of time reversal, in turn, will change the sign of M, while the sign of P remains invariant. Therefore non-magnetic ferroelastics and ferroelectrics are invariant under time reversal whereas ferromagnets and ferrotoroids are not. Magnetoelectric multiferroics require simultaneous violation of space and time inversion symmetry since they are both ferromagnetic and ferroelectric.

A crystal can be classified into 32 point groups. In addition, with the consideration of magnetic symmetry, 90 magnetic point groups can be distinguished. Therefore, as a whole, there are 122 point groups (also known as Shubnikov-Heesch magnetic point groups) including non-magnetic point groups [Landau et al., 1984]. Among 122 magnetic point groups, there are 31 allowing a spontaneous magnetization and 31 allowing a spontaneous polarization. However, only 13 point groups allow both spontaneous magnetization and electric polarization to occur [Schmid, 1994, Birss, 1964],and many materials occur in one of these 13 point groups without being multiferroic.

Table 1.1 Ferroics-symmetry

Symmetry	Space invariant	Space variant
Time invariant	Ferroelastic	Ferroelectric
Time variant	Ferromagnetic	Ferrotoroidic

1.2.4   Classification of multiferroics

Single-phase multiferroic materials can be classified into two big groups and other subgroups, according to the origin of ferroelectric ordering in them [Khomskii, 2009].

Type I multiferroics:

This group of materials is the most numerous and in these materials, the ferroelectricity and magnetism arise from different mechanisms and occur at different temperatures. Usually, the ferroelectricity appears at high temperature, and the magnetic ordering, which is usually antiferromagnetic, sets in at low temperature. This group canbe further divided into three subgroups: i) Ferroelectricity due to lone paired electrons, ii) Ferroelectricity due to charge ordering and iii) Geometrically driven ferroelectricity.

i) Ferroelectricity due to lone paired electrons:

In the group of perovskite materials with the ABO_3 formula (also in $ABB'O_3$), the ferroelectric displacement is induced by the cation on the A-site and the magnetism arises from the partially filled d shell cation on the B-site. In these materials, A-site cation (i.e. Bi^{3+}, Pb^{3+}, ...) has active $6s^2$ lone-pair of electrons, and off-centering of the A-site cation is caused by energy–lowering electron sharing between the formally empty A-site (6p) orbital and the filled O (2p) orbital [Waghmar et al., 2003]. Examples of this group are $BiFeO_3$, $PbVO_3$, $Bi_{12}NiMnO_6$ [Wang et al., 2003, Sakai et al., 2007].

ii) Ferroelectricity due to charge ordering:

Ferroelectricity can be caused by charge ordering in compounds containing transition metal ions with mixed-valence and magnetic frustration. These metal ions form a polar arrangement, causing ferroelectricity. If magnetic ions are present, a coexisting magnetic order can be originated and may be coupled to ferroelectricity. Typical examples of this group are $LuFe_2O_4$ [Hur et al., 2004], (Pr, Ca) MnO_3 [Brink and Khomskii, 2008].

iii) Geometrically driven ferroelectricity:

In some hexagonal manganite compounds $RMnO_3$ (R=Y or small rare earth), the tilting of the rigid MnO_5 block generates a closer packing in the perovskite. As a result, there is a displacement of oxygen ions towards the A-site ions, thus, causes ferroelectricity. Examples are $YMnO_3$, $LuMnO_3$ [Van Aken et al., 2003].

Type II multiferroics:

This group is distinguished by a strong coupling between magnetism and ferroelectricity where magnetism drives ferroelectricity. In this case, ferroelectricity sets in at the same temperature as magnetic ordering and is driven by it. It can be classified also into two subgroups: i) Spiral magnets and ii) Collinear magnets.

i) Spiral magnets:

In most oxide materials, ferroelectricity originates in conjunction with a spiraling magnetic phase, mostly of the cycloid type. In these materials, the induced polarization lies in the plane in which the cycloid resides and is orthogonal to its propagation vector. An abrupt change of the cycloidal spiral plane induced by the magnetic field results in the corresponding rotation of the polarization vector. This magnetoelectric coupling is due to the so-called inverse Dzyloshinsky-Morriya interaction,

$$P = A\Sigma r_{ij} \times (S_i \times S_j) \qquad\qquad .... (1.2.5)$$

where, A is spin-orbit interaction and r_{ij} is the vector connecting the spins S_i and S_j [Eerenstein et al., 2006]. Most of the type-II multiferroics known up-to-date belong to this category. Examples of such materials are $TbMnO_3$, $DyMnO_3$ [Kimura et al., 2003].

ii) Collinear magnets:

In some materials having strong uniaxial isotropy, all magnetic moments are aligned along a particular axis and can stabilize a periodic collinear spin arrangement of the type (↑↑↓↓) up-up-down-down without involving the spin-orbit interaction. The distortion of the ferro (↑↑) and antiferro(↑↓) bonds is different due to an exchange striction and hence the material becomes ferroelectric, because of the ordered dipoles. The simplest examples of this group are ortho-$HoMnO_3$, Ca_3CoMnO_6 [Lee et al., 2011].

Lanthanum orthoferrite ($LaFeO_3$) could be classified as type II [Mukhopadhyay et al., 2013] spiral multiferroic [Das et al., 2007, Zhang et al., 2006] as the magnetism causes ferroelectricity in this material.

1.3 Crystal structure, magnetic, electric and magnetoelectric properties of $LaFeO_3$

1.3.1 Crystal structure

The most common structure for multiferroic materials is the perovskite with ABO_3 formula, where, A and B are cations. $LaFeO_3$ is one of the well-known perovskite structures exhibiting multiferroic properties. It crystallizes in an orthorhombically distorted perovskite structure, which is deviated from the ideal cubic perovskite structure through the distortion of the FeO_6 octahedra [Liu et al., 2002, Hill, 2000]. The structural distortion in $LaFeO_3$ is due to the mismatch between the size of La and Fe cations. In the orthorhombic perovskite structure of $LaFeO_3$, each unit cell has corner linked octahedra FeO_6, in which the Fe atom is surrounded by six O^{2-} ions. The O1 atom is at the apical position and the O2 atom is at the equatorial position of the octahedron. The La atom occupies the space between the FeO_6 octahedra [Liu et al., 2002, Hill, 2000]. The orthorhombic unit cell of $LaFeO_3$ is shown in Figure 1.3(a).

The steadiness and distortion of the perovskite structure ABO_3 (where, A is a rare earth element and B is a transition metal) depend on the tolerance factor "t" and is determined by the Goldschmidt tolerance factor:

$$t = (R_A + R_O)/ \sqrt{2} (R_B + R_O) \quad \ldots\ldots \quad (1.3.1)$$

where, R_A, R_B, and R_O are the radii of A-cation, B-cation and O-anion respectively [Goldschmidt, 1926]. If t = 1, then the perovskite adopts a cubic structure [Rick, 2007]. If t > 1, the crystal structure becomes hexagonal. If 0.96 < t < 1, the perovskite assumes a rhombohedral structure. If t < 0.96, the perovskite distorts into an orthorhombic structure. For $LaFeO_3$, t is less than 1. Therefore, the cubic structure transforms into the orthorhombic one and leads to the deviation of the Fe–O–Fe bond angle from 180°. This divergence in Fe–O–Fe bonds further leads to a distortion in FeO_6 octahedra.

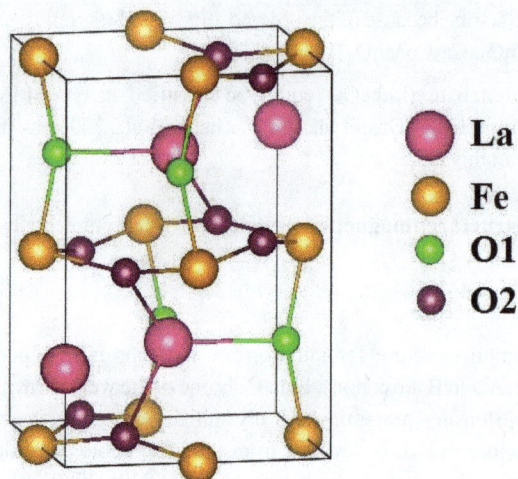

Figure 1.3(a) Orthothombic unit cell of LaFeO₃

1.3.2   Magnetic properties

$LaFeO_3$ exhibits G-type antiferromagnetic behavior with a high Néel temperature (T_N) of 740 °C [Hearne and Pasternak, 1995]. The G-type ordering observed on the simple cubic lattice of $LaFeO_3$ antiferromagnet [Koehler et al., 1960] is given in Figure 1.3(b). Figure 1.3(b) shows both intra-plane and inter-plane coupling are antiferromagnetic. The antiferromagnetic coupling in $LaFeO_3$ is due to the collinear antiferromagnetic (AFM) spin order caused by the super-exchange interaction between the magnetic cations (Fe^{3+}) via oxygen (O^{2-}) anion [Anderson, 1950]. The super-exchange interaction depends on the covalent bonding of the metal ions Fe^{3+} with their bridging anion (O^{2-}). If Fe^{3+}–O^{2-}–Fe^{3+} pathway has a linear bond angle of 180°, as in Figure [1.3(c)](a) [Anderson, 1950], then the strongest interactions are expected to be antiferromagnetic. An excited state is formed in which a double charge transfer occurs between oxygen and iron sites with the net effect being that neighboring Fe^{3+} ions are coupled antiparallel. However, the ferromagnetic ground state can be stabilized by super- exchange interaction when the Fe^{3+}–O^2–Fe^{3+} pathway is at 90°, as illustrated in Figure [1.3(c)](b) [Anderson, 1950], which arises as a result of mutually perpendicular oxygen orbital in the exchange pathway. After the double oxygen to iron charge transfer, the ferromagnetic excited state on the orthogonal oxygen orbital is lower in energy than the antiferromagnetic excited state due to Hund's rule. Hence, the strength of the exchange coupling via oxygen is highly dependent on the Fe^{3+}–O^2–Fe^{3+} bond angle. The transition from antiferromagnetic to ferromagnetic state occurs at ~96° [Anderson, 1950]. In $LaFeO_3$, La^{3+} is non-magnetic as its valence electrons are in pairs and so the magnetic moments of Fe^{3+} and the interactions between and Fe^{3+} and La^{3+} ions are the primary sources of the magnetic properties. The other factors influencing the magnetization of $LaFeO_3$ are (i) the difference between the magnetic moment of Fe^{3+} (5 µB) and Fe^{2+} (4 µB) ions, (ii) the direct magnetic interaction between nearest neighbours of Fe^{3+} moments and (iii) the coupling of spins through the oxygen ions due to super-exchange interaction.

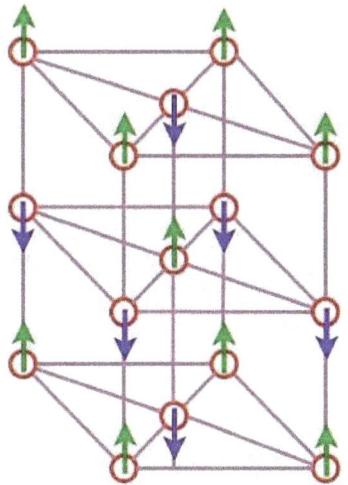

Figure 1.3(b) *G-type antiferromagnetic ordering on simple cubic Lattice of LaFeO₃*

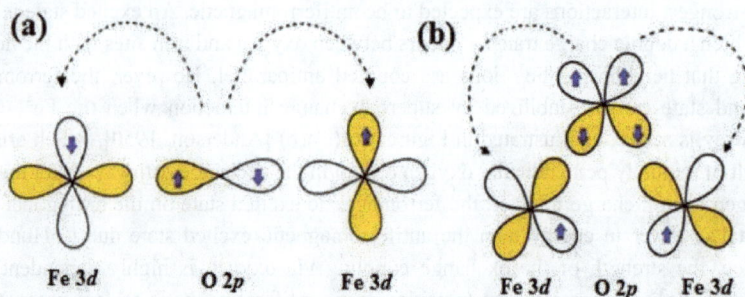

Figure 1.3(c) (a) *A simple view of super-exchange in a linear Fe–O–Fe system giving antiferromagnetic alignment of magnetic iron ions with a bridging oxygen anion, **(b)** A simple view of the super-exchange pathway stabilizing ferromagnetic coupling in a 90° Fe–O–Fe array*

1.3.3 Electric properties

1.3.3.1 Dielectric mechanism in LaFeO₃

Polarization in dielectrics is the alignment of electric dipole moments in the presence of an applied electric field. The polarization in dielectric material occurs due to various microscopic mechanisms and they are classified as, atomic, ionic, orientation and space charge polarization [Moulson and Herbert, 2003].

i.   Atomic polarization is due to the relative shift of negative charges (electron cloud) and positively charged nucleus of the atom in the presence of an external electric field.

ii. Ionic polarization occurs due to the relative displacement of cations and anions in the presence of an applied electric field.

iii. Orientational polarization is caused by the rotation of permanent dipole momentsinto an applied electric field direction.

iv. Space charge polarization arises due to the displacement of charge carriers by an applied electric field and is stopped at the surfaces or grain boundaries of the material. This displacement of charge carriers to the surface creates electric dipoles.

LaFeO₃ is a good dielectric material due to the space charge polarization. In this material, the localized charges i.e., electrons and holes can hop from one crystallographic site to another, and create the hopping polarization. The presence of $La^{3+} \leftrightarrow La^{4+}$ ions gives rise

Materials Research Forum LLC
https://doi.org/10.21741/9781644902271

to p-type carriers and their local displacement in the applied electric field direction contributes to the net polarization in addition to n-type carriers. However, the contribution of holes (p-type charge carriers) from $La^{3+} \leftrightarrow La^{4+}$ ions is smaller than that of electrons (n-type charge carriers) from $Fe^{2+} \leftrightarrow Fe^{3+}$ ions and opposite in sign [Ponpian et al., 2002]. $LaFeO_3$ exhibits normal dielectric dispersion as shown in frequency- dependent dielectric constant plot Figure 1.3(d). Figure 1.3(d) shows the dielectric constant decreases steeply at low frequencies and remains constant at high frequencies. This behavior is attributed to the dipoles resulting from the change in the valence state of cations and space charge polarization. At high frequencies, the dielectric constant remainsindependent of frequency due to the inability of $Fe^{2+} \leftrightarrow Fe^{3+}$ electric dipoles to follow the applied alternating electric field [Patankar et al., 2001]. The high dielectric constant at low frequencies is associated with the small polaron hopping mechanism, which results in electronic polarization contributing to low-frequency dispersion. This is also attributed to Maxwell–Wagner [Maxwell, 1973, Wagner, 1993] type interfacial polarization which is in agreement with Koop's theory [Koops, 1951].

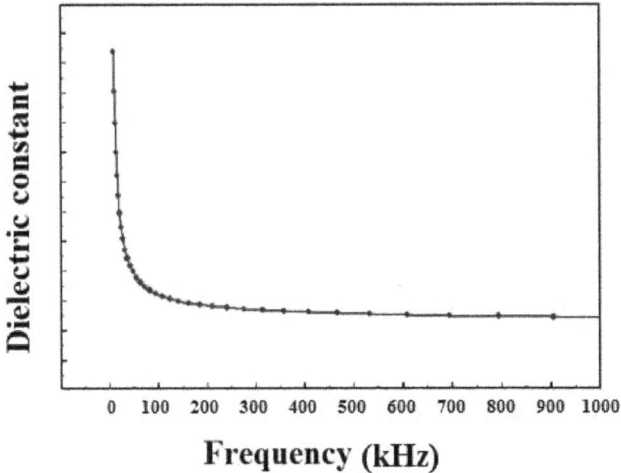

Figure 1.3(d) Frequency-dependent dielectric constant plot of $LaFeO_3$

1.3.3.2 Electrical conduction mechanism

The electrical conduction of $LaFeO_3$ multiferroic can be related to small polarons [Adler and Feinleib, 1970] as its conductivity increases with frequency. The conductivity in this material is due to both n-type and p-type charge carriers. The ac conductivity (σ_{ac}) can be estimated from the dielectric data by the relation [Chougule and Chougule, 2007].

$$\sigma_{ac} = 2\pi\varepsilon_0 \varepsilon' f \tan \delta \qquad\qquad \text{......... (1.3.2)}$$

where, ε_0 is the permittivity of free space ($\varepsilon_0 = 8.854 \times 10^{-12}$ F/m), ε' is the dielectric constant of the sample, f is the frequency of the applied field and $\tan \delta$ is the dielectric loss factor (dielectric loss tangent). The frequency-dependent ac conductivity plot, which is illustrated in Figure 1.3(e) shows the increasing trend in conductivity with increasing applied frequency. At low frequencies, the conductivity remains constant due to the inability of Fe^{2+} and Fe^{3+} ions to follow the applied electric field. At high frequencies, the conductivity increases, because of the effect of the applied electric field that promotes the hopping or transfer of charge carriers between localized $Fe^{2+} \leftrightarrow Fe^{3+}$ ions and $La^{3+} \leftrightarrow La^{4+}$ ions states [Vishwanathan and Murthy, 1990].

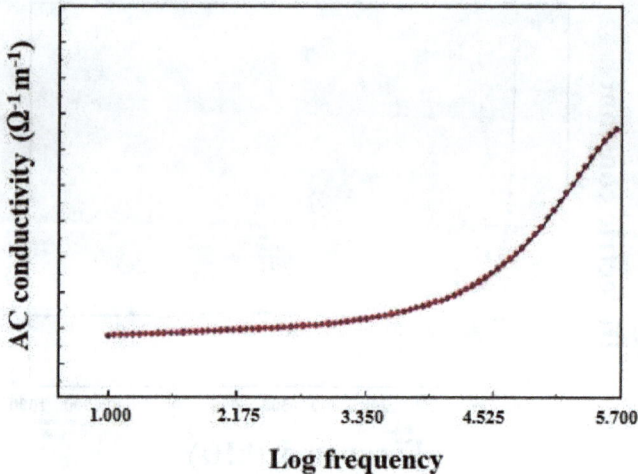

Figure 1.3(e) *Frequency-dependent ac conductivity plot of $LaFeO_3$*

1.3.3.3 Ferroelectric properties

In $LaFeO_3$, ferroelectric ordering is caused by magnetic ordering. In this system, FeO_6 octahedra rotate due to the low symmetry of the distorted orthorhombic perovskite- type structure, while the Fe^{3+} cation slightly shifts from the centrosymmetric position. Therefore, the displacement of the La^{3+} ion with FeO_6 octahedra causes ferroelectricity and Fe-O-Fe super-exchange interaction leads to the canted AFM ordering in the system at T_N. Thus, in $LaFeO_3$, magnetism causes ferroelectricity at Néel temperature (T_N). The electric polarization in this system is caused by the rotational displacement of $Fe^{2+} \leftrightarrow Fe^{3+}$ dipoles (*n*-type charge carriers) and the local displacement of $La^{3+} \leftrightarrow La^{4+}$ dipoles (*p*-type charge carriers) [Thirumalairajan et al., 2015]. In general, ferroelectricity is harder to demonstrate in polycrystalline materials like ceramics than in a single crystal, which is due to the random orientation of crystallites. Consequently, in some single crystals, the polarization reverses quite abruptly to form a square loop, as shown in Figure[1.3(f)](a), while in most ceramics the loop is rounded, as shown in Figure [1.3(f)](b) [Vandeven et al., 1967].

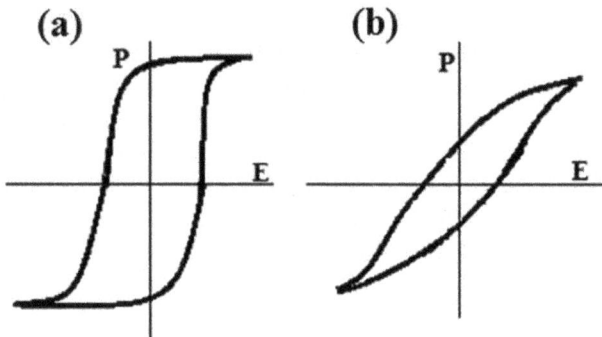

Figure 1.3(f) Schematic diagrams of P-E loop for (a) single crystal, (b) polycrystalline

1.3.4 Magnetoelectric properties

$LaFeO_3$ is a canted G-type wide-gapped antiferromagnetic insulator with high Néel temperature ($T_N \sim 740\ °C$) [Hearne and Pasternak, 1995]. It exhibits a strong coupling between magnetism and ferroelectricity, where magnetism causes ferroelectricity. The 3d electrons of transition metal ion Fe^{3+}, which are responsible for magnetic ordering in $LaFeO_3$, induce a lattice distortion in the system. This distortion creates a strong local electric field in the host matrix, which is responsible for the onset of ferroelectric ordering in the system. Thus, in this material, ferroelectricity sets in at the same temperature as magnetic ordering (i.e., at Néel temperature (T_N)) and the magnetic ordering is due to cycloidal spin structure. Therefore, it could be classified as type II- spiral multiferroic. [Mukhopadhyay et al., 2013, Das et al., 2007, Zhang et al., 2006].

1.4 Literature review on pure and substituted lanthanum orthoferrites

1.4.1 Magnetoelectric properties of pure and substituted LaFeO$_3$

The synthesis method and the substitution level of various cations at the La^{3+} site and/or Fe^{3+} site in LaFeO$_3$ modify greatly the magnetoelectric properties of LaFeO$_3$. In an attempt to get improved/modified magnetoelectric properties, there are a few reports on pure and substituted LaFeO$_3$.

Ahmed et al., [2015] have reported that Bi-substituted LaFeO$_3$ (La$_{1-x}$Bi$_x$FeO$_3$; $0 \leq x \leq 0.2$) exhibits weak ferromagnetic behavior. The value of saturation magnetization (M_s) increases with increasing Bi content up to x= 0.10 and then decreases, while the dielectric constant and ac conductivity decrease and then increase.

Mukhopadhyay et al., [2013] synthesized nanostructured La$_{1-x}$Zn$_x$FeO$_3$ (x=0, 0.1and 0.3) by chemical co-precipitation method. They have observed the exchange bias (EB) effect and enhanced soft magnetic property with Zn substitution. They have reported the frequency dependence of dielectric constant for the samples. Also, they have reported that the values of ferroelectric parameters such as P_m, P_r and E_c for LaFeO$_3$ are larger than those values observed for Zn-substituted samples. The exchange bias effect, exchange anisotropy becomes very prominent in 30% Zn-substituted LaFeO$_3$ (La$_{0.7}$Zn$_{0.3}$FeO$_3$).

The magnetic and dielectric properties of nano-crystalline LaFe$_{1-x}$Zn$_x$O$_3$ ($0 \leq x \leq 0.3$) synthesized by sol-gel auto-combustion technique have been studied by Bhat et al., [2013]. They have observed weak ferromagnetic ordering and a shift of magnetic hysteresis curves to the negative direction (negative EB effect) of the field. They have reported that doping of Zn^{2+} greatly enhances the dielectric constant and it shows a colossal value.

Triyono et al., [2018] have reported the structure and dielectric properties of Sr-substituted LaFeO$_3$ (La$_{1-x}$Sr$_x$FeO$_3$ (x= 0.1, 0.2, 0.3 and 0.4)) prepared by sol-gel and sintering methods. They have observed that Sr-substitution induces the concurrent creation of Fe^{4+} and Fe^{5+} to maintain the charge neutrality in the crystal lattice and also increases the dielectric constant and electronic conductivity.

Nguyen et al., [2014] synthesized nano La$_{1-x}$Sr$_x$FeO$_3$ perovskite materials (x=0.0, 0.1, 0.2 and 0.3) using the co-precipitation method and have studied their magnetic properties.

They have reported that the increase of Sr^{2+} doping increases the coercive force of synthesized $La_{1-x}Sr_xFeO_3$.

Ti-substituted $LaFeO_3$ ($LaFe_{1-x}Ti_xO_3$, x=0, 0.1 and 0.2) were synthesized by Phokha et al., [2015] using polymer pyrolysis method. They have observed weak ferromagnetic behavior for all samples with a maximum magnetization of 0.32 emu/g for x=0.2. All the samples exhibit a giant dielectric constant of $\sim 8 \times 10^3 - 18 \times 10^3$.

Acharya [Acharya and Chakrabarti, 2010] have reported that for Al-substituted $LaFeO_3$ ($La_{0.5}Al_{0.5}FeO_3$) prepared by solid state reaction method, the magnetization and magnetic susceptibility are considerably enhanced compared with $LaFeO_3$. The frequency-dependent dielectric constant and temperature-dependent dielectric constant predict the existence of spontaneous polarization in the sample. Similarly, Al^{3+} substitution results in ferromagnetism in $La_{1-x}Al_xFeO_3$ (x=0.0, 0.05, 0.1, 0.2, 0.3, 0.4 and 0.5) nanopowders prepared by polymerization complex method [Janbutrach et al., 2014] and also enhances ferromagnetic property of Al-substituted samples.

The effect of Ce^{3+} doping on the magnetic property of $La_{1-x}Ce_xFeO_3$ prepared using solution combustion method has been reported by Shikha et al., [2015]. The magnetic properties like saturation magnetization and remanent magnetization have been greatly enhanced by Ce^{3+} doping, where a small amount of substituted Ce^{3+} may be oxidized to Ce^{4+}, thereby producing a remarkable effect on physicochemical properties. No dielectric and ferroelectric studies have been reported for $La_{1-x}Ce_xFeO_3$.

1.4.2 Applications of pure and substituted lanthanum orthoferrites

1.4.2.1 Applications of pure LaFeO₃

$LaFeO_3$ has the potential for controlling magnetism using electric fields via their magnetoelectric coupling; it is used for developing magnetoelectronic devices. These include novel spintronic devices [Gajek et al., 2007, Nan et al., 2008] such as

i.   Tunnel magnetoresistance (TMR) sensors

ii. Spin valves with electric field tunable functions

iii. High-sensitivity ac magnetic field sensors

iv. Electrically tunable microwave devices

Due to the co-existence of coupled ferroic ordering, these materials have also been proposed for various applications in non-volatile magnetic memory devices, and ultrasensitive magnetic read heads of modern hard disk drives [Moser et al., 2000, Parkin et al., 1999]. Furthermore, it can be used in the construction of multistate memories in which data are written electrically and read magnetically [Hill, 2000].

Beyond magnetoelectricity-based applications, there are several other applications of $LaFeO_3$ because it has various interesting properties. It has high electrical conductivity, excellent thermal stability, so it can be used as separator material in solid oxide fuel cell (SOFC) [Steele and Heinzel, 2001, Devi et al., 2004], as a hot electrode for magnetohydrodynamic (MHD) power generation [Vasques et al., 1998], catalytic activity in the complex oxidation of hydrocarbons and catalytic combustion of methane [Meadowcraft and Wimmer, 1979]. $LaFeO_3$ is used as a sensor to detect acetone and other gases [Liu et al., 2006, Toa, et al., 2003] due to its structural defects caused by fluctuation between two stable oxidation states of iron.

1.4.2.2 Applications of substituted $LaFeO_3$

Sr-substituted $LaFeO_3$ can be used as an ethanol gas sensor with low power consumption [Liu et al., 2007]. Due to mixed electron and oxygen ion conductivity, Sr- substituted $LaFeO_3$ can be used as the main constituent of inexpensive and efficient ceramic membranes for oxygen separation and primary processing of natural gas [Kozhevnikov et al., 2009, Kharton, 2011]. Sr-substituted $LaFeO_3$ show hard magnetic property with high coercive force (Hc > 15.92 kA/m), therefore they can be used in the production of permanent magnets for motors [Nguyen et al., 2014]. Pristine and Zn- substituted $LaFeO_3$ multiferroics exhibit exchange bias (EB), so that they can be used in spin-valve technology for controlling magnetization in devices such as giant magnetoresistance sensors, magnetic random access memories and switching devices [Wolf et al., 2001, Parkin et al.,1999, Mukhopadhyay et al., 2013]. The schematic diagram of the magnetoelectric random access memory (ME RAM) cell is shown in Figure 1.4(a). As Al-substituted $LaFeO_3$ exhibit ferromagnetism with enhanced magnetic properties they can be utilized for magnetic memory device applications [Janbutrach etal., 2014]. Ce-substituted $LaFeO_3$ can absorb visible light therefore they can be exploited for their possible application as visible-light-driven photocatalyst [Shikha et al., 2015].

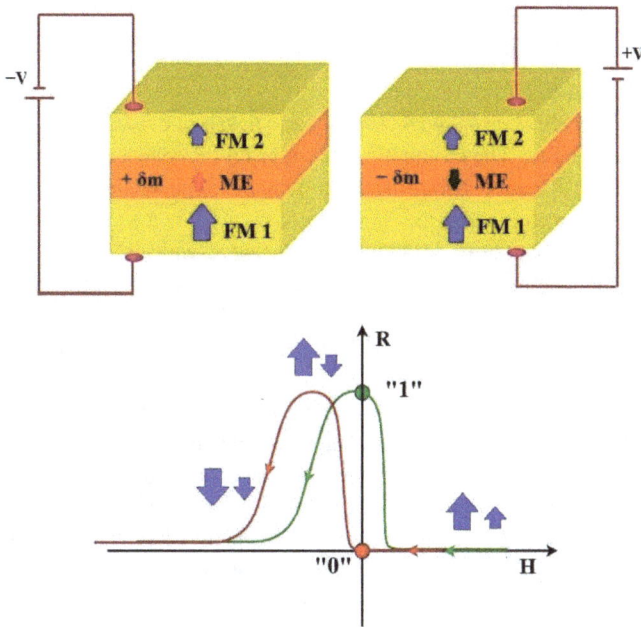

Figure 1.4(a) *Magnetoelectric random access memory (ME RAM) cell in its two switching stages (i) "0" (ii) "1"*

1.5 Synthesis of lanthanum orthoferrite-type multiferroics

1.5.1 Solid state reaction method

Literature reveals that various techniques have been employed to synthesize pure and substituted $LaFeO_3$ such as sol-gel method [Xiang et al., 2013], micro-emulsion method [Giannakas et al., 2006], chemical co-precipitation method [Shikha et al., 2015], hydrothermal method [Xiao et al., 2015] microwave-assisted method [Tang et al., 2013], solid state reaction method [Acharya et al., 2010], etc. Among these methods, solid state reaction (SSR) method is widely used because it eliminates intermediate impurity phases and therefore products with pure phase can be obtained. This method has higher productivity and lower cost. Furthermore, high temperature sintering, which is being followed in this method, increases the rate of diffusion and all samples of varying concentrations are synthesized at the same time under the same environment. Above all, it is environmental friendly because there is no need for solvents and also no need for purification after synthesis.

In solid state reaction (SSR) method, the starting chemicals are weighed accordingto the desired stoichiometry and mixed thoroughly, and then heat treated at high temperature with intermediate grinding. The essential steps needed for the solid state reaction technique are (i) mixing, (ii) calcination, (iii) grinding, (iv) pressing / pelletizing and (v) final sintering. The flow chart which explains the various steps involved in a typical solid state reaction is presented in Figure 1.5(a). Initially, the starting chemicals which are taken in the appropriate stoichiometric ratios are mixed thoroughly to get a homogenous mixture of reactants. Then, the mixed precursor powders are calcined at hightemperature to enhance the inter diffusion of their ions and hence the interaction between them. Calcination is the pre-sintering process in which a material is heated at a temperature, lesser than the melting point of the end product material. Calcination also promotes in homogenizing the materials and controls their shrinkage in the subsequent sintering process. The calcined materials are then thoroughly ground for several hours to homogenize the compositional variations, which may still exist during calcination. To increase the intimate contact of crystallites and thereby the reaction rate, pelletizing the powders is done before sintering which is important in SSR method. Sintering of pelletized powder samples is done at a high temperature below the melting point of the samples, and so the crucibles (sample containers) must be able to tolerate high temperature. In the sintering process, atomic diffusion occurs and therefore the particles fuse together and form high dense ceramics as the end product. It is known that the purity of the raw chemicals, degree of mixing, calcination and sintering time and their temperature, grain size and shape, dopant concentrations, etc., greatly influence the physical properties of the end product materials. Therefore, during synthesis, the above said factors have to be monitored and controlled to produce suitable materials for technological applications.

In this work, lanthanum orthoferrite-type multiferroics have been synthesized using the solid state reaction technique. The precursor materials were weighed using a digital balance (Model MK 200E) with a readability of 0.001 gm. For mixing and grinding the starting materials effectively, the agate mortar with 12 cm in diameter (Figure [1.5(b)](a)) was used. For pelletizing the ground powder samples, a hydraulic pelletizer (Figure [1.5(b)] (b)), which can provide a maximum pressure of 15 tons was used. To make the samples as disc-shaped pellets, a die with a 1.2 cm diameter was used. In this work, alumina crucibles (boat type) (Figure [1.5(b)](c)), which withstand a temperature of 1650 °C were used as sample containers. For calcination and sintering processes, a high temperature tubular furnace (Figure 1.5(c)) equipped with Nippon/Eurotherm PID programmable controller was used. This programmable tubular furnace has a working temperature limit up to 1600

°C with 1 °C accuracy of dwell temperature and has a heating and cooling rate of 1 °C/min to 5 °C/min.

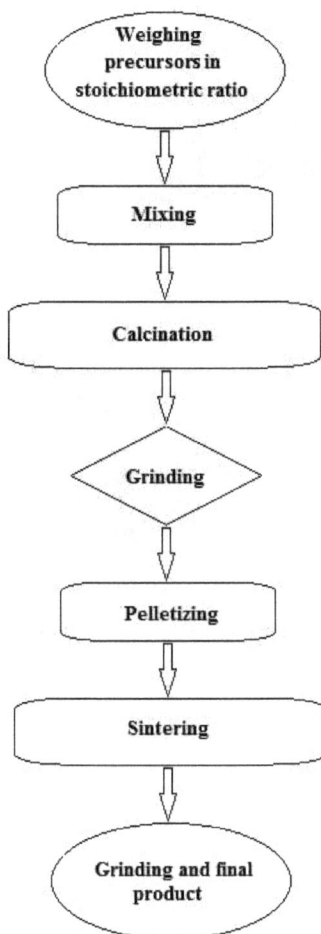

Figure 1.5(a) *Flow chart of a typical solid state reaction technique*

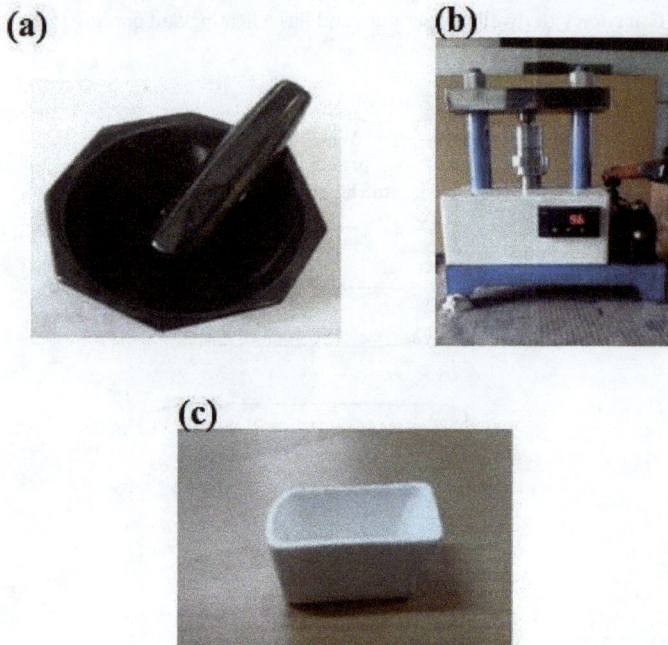

(a)

(b)

(c)

Figure 1.5(b) *Photographs of **(a)** Agate mortar and pestle, **(b)** Pelletizer, **(c)** Alumina crucible*

Figure 1.5(c) *Photograph of high temperature tubular furnace (temperature range up to 1600 °C)*

1.5.2 Synthesis of lanthanum orthoferrite–type multiferroics

In the present work, four different series of lanthanum orthoferrite-type multiferroics viz., $La_{1-x}Ce_xFeO_3$, $La_{1-x}Zn_xFeO_3$, $La_{1-x}Al_xFeO_3$ and $La_{1-x}Sr_xFeO_3$ have been synthesized by solid state reaction method. High purity metal oxide powders ($\geq$ 99.99%, Alfa aesar) were taken according to the stoichiometric composition and were used as precursor materials. The molecular weights of the precursor materials used to synthesize LFO-type multiferroics are tabulated in Table 1.2.

Table 1.2 *Molecular weights of precursor chemicals used for sample preparation*

Chemical	Molecular Weight (gm/mole)
La_2O_3	325.81
CeO_2	172.115
ZnO	81.408
Al_2O_3	101.96
$SrCO_3$	147.63
Fe2O3	159.69

1.5.2.1 Ce-substituted lanthanum orthoferrites − $La_{1-x}Ce_xFeO_3$

To synthesis cerium substituted lanthanum orthoferrite multiferroics $La_{1-x}Ce_xFeO_3$(x=0.00, 0.03, 0.06, 0.09 and 0.12), high purity oxides lanthanum oxide (La_2O_3, 99.99 %), cerium oxide (CeO_2, 99.99 %) and iron (III) oxide (Fe_2O_3, 99.999%) were used as precursor materials. The compositions of the oxides were taken according to the chemical reaction equation given below:

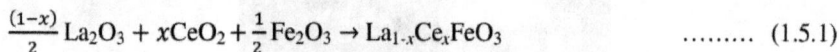

$$\frac{(1-x)}{2} La_2O_3 + xCeO_2 + \frac{1}{2} Fe_2O_3 \rightarrow La_{1-x}Ce_xFeO_3 \qquad \dots\dots\dots (1.5.1)$$

Stoichiometric quantities of precursor materials used to synthesis $La_{1-x}Ce_xFeO_3$ are presented in Table 1.3. Initially, the metal oxide powders were mixed thoroughly in an agate mortar for 4 h to produce a fine mixture. Then, the ground powders were placed in alumina crucibles and calcined in air atmosphere at 1000 °C for 10 h using a tubular furnace. The calcined powders were ground for 10 h and disc-shaped pellets of 12 mm diameter and ~1mm thickness were made from the ground mixture using a pelletizer by applying hydraulic pressure of 6 tons. Then, the pellets were heat treated in air at 1300 °C for 16 h with a heating rate of 5 °C/min. The heat treated pellets were again ground for another 6 h and pelletized and were sintered out in air atmosphere at 1300 °C for 16 h. The sintered $La_{1-x}Ce_xFeO_3$ (x=0.00, 0.03, 0.06, 0.09 and 0.12) pellets are shown in Figure 1.5(d).

Table 1.3 *Quantities of precursor chemicals used for the synthesis of $La_{1-x}Ce_xFeO_3$*

	Weight (gm)		
Ce concentration	La_2O_3	CeO_2	Fe_2O_3
0.00	3.620	−	1.774
0.03	3.511	0.114	1.774
0.06	3.402	0.229	1.774
0.09	3.294	0.344	1.774
0.12	3.185	0.458	1.774

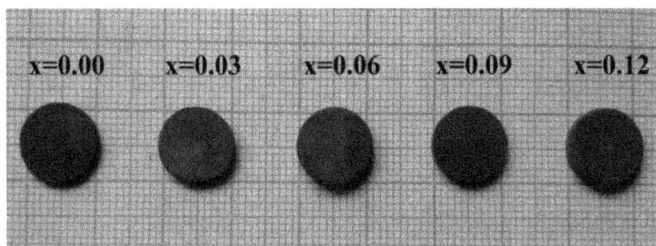

Figure 1.5(d) Sintered pellets of $La_{1-x}Ce_xFeO_3$ (x=0.00, 0.03, 0.06, 0.09 & 0.12)

1.5.2.2 Zn-substituted lanthanum orthoferrites - $La_{1-x}Zn_xFeO_3$

The divalent element zinc substituted lanthanum orthoferrite multiferroics $La_{1-x}Zn_xFeO_3$ (x=0.00, 0.05, 0.15, and 0.25) were prepared using high purity metal oxide powders, lanthanum oxide (La_2O_3, 99.99 %), Zinc oxide (ZnO, 99.99 %) and iron (III) oxide (Fe_2O_3, 99.999 %) as starting raw materials. To synthesis $La_{1-x}Zn_xFeO_3$, the compositions of constituent metal oxides were taken according to the following chemical reaction equation.

Table 1.4 presents the quantities of metal oxides used as precursor materials for the synthesis of $La_{1-x}Zn_xFeO_3$. As the first step, stoichiometrically weighed oxide powders

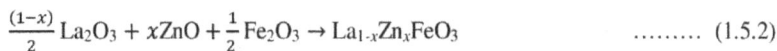

$$\frac{(1-x)}{2} La_2O_3 + xZnO + \frac{1}{2} Fe_2O_3 \rightarrow La_{1-x}Zn_xFeO_3 \qquad \text{......... (1.5.2)}$$

were mixed well by using an agate pestle mortar for 4 h to produce a homogenous mixture of reactants. Then, the mixed powders were placed in alumina crucibles which can withstand a temperature of 1650 °C and calcined in air at 1000 °C for 10 h with a heating rate of 5 °C/min, using the programmable electric furnace. The calcined powders were ground for 10 h and were heat treated in air at 1200 °C for 12 h. The heat treated powders were ground for another 10 h and circular pellets were made from them by applying a hydraulic press of 6 tons. Then the circular pellets were finally sintered out in air at 1300 °C for 16 h. The sintered $La_{1-x}Zn_xFeO_3$ (x=0.00, 0.05,0.15 and 0.25) pellets are shown in Figure 1.5(e).

Table 1.4 *Quantities of precursor chemicals used for the synthesis of* $La_{1-x}Zn_xFeO_3$

	Weight (gm)		
Zn concentration	**La_2O_3**	**ZnO**	**Fe_2O_3**
0.00	3.620	-	1.774
0.05	3.439	0.090	1.774
0.15	3.220	0.283	1.856
0.25	2.909	0.485	1.901

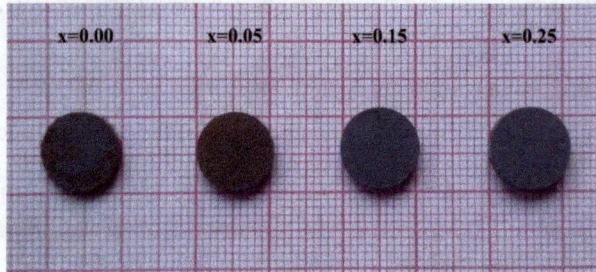

Figure 1.5(e) *Sintered pellets of* $La_{1-x}Zn_xFeO_3$ *(x=0.00, 0.05, 0.15 & 0.25)*

1.5.2.3 Al-substituted lanthanum orthoferrites − $La_{1-x}Al_xFeO_3$

Aluminium substituted $La_{1-x}Al_xFeO_3$ (x=0.05, 0.15 and 0.25) multiferroics were synthesized using highly pure reagents lanthanum oxide (La_2O_3, 99.99%), aluminium oxide (Al_2O_3, 99.99%) and iron (III) oxide (Fe_2O_3, 99.99%). The compositions of reagents were taken in accordance with the following chemical reaction equation.

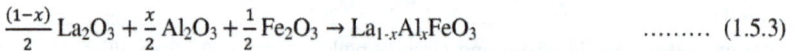

$$\frac{(1-x)}{2}\, La_2O_3 + \frac{x}{2}\, Al_2O_3 + \frac{1}{2}\, Fe_2O_3 \rightarrow La_{1-x}Al_xFeO_3 \qquad \dots\dots\dots (1.5.3)$$

The starting reagents were weighed in their stoichiometric ratio using a digital balance which has a readability of 0.001 gm. The quantities of chemicals used to synthesis $La_{1-x}Al_xFeO_3$ are given in Table 1.5. The weighed powders were mixed manually for 4 h using an agate mortar and pestle. The mixed powders were placed in alumina crucibles, then covered with a lid and calcined in air twice at temperature 1000 °C for 10 h, with a heating rate of 5 °C/min, using a programmable tubular electric furnace with intermediate grinding for 10 h. The calcined powders were reground thoroughly for 10 h and then compressed in the form of dense circular pellets by applying hydraulic pressure of 6 tons. Then, the final sintering of the pellets was done at 1300 °C for 16 h. The sintered pellets of $La_{1-x}Al_xFeO_3$ (x=0.05, 0.15 and 0.25) multiferroics prepared through solid state reaction method are shown in Figure 1.5(f).

Table 1.5 *Quantities of precursor chemicals used for the synthesis of* $La_{1-x}Al_xFeO_3$

Al concentration	Weight (gm)		
	La_2O_3	Al_2O_3	Fe_2O_3
0.05	3.439	0.056	1.774
0.15	3.296	0.182	1.901
0.25	3.054	0.318	1.996

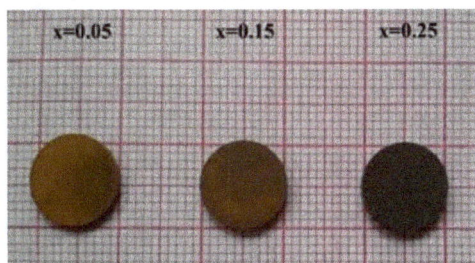

Figure 1.5(f) *Sintered pellets of* $La_{1-x}Al_xFeO_3$ *(x=0.05, 0.15 & 0.25)*

1.5.2.4 Sr-substituted lanthanum orthoferrites - $La_{1-x}Sr_xFeO_3$

Strontium substituted lanthanum orthoferrite multiferroics ($La_{1-x}Sr_xFeO_3$) with different Sr concentrations (x=0.05, 0.10, 0.15 and 0.20) were synthesized by conventional high temperature solid state reaction method. High purity analytical grade chemicals lanthanum oxide (La_2O_3, 99.99%), strontium carbonate ($SrCO_3$, 99.99%) and iron (III) oxide (Fe_2O_3, 99.999%) were used as raw materials. The compositions of chemicals were weighed using a digital balance according to the following chemical reaction equation. The quantities of chemicals used for the synthesis of $La_{1-x}Sr_xFeO_3$ are given in Table 1.6. Initially, the raw chemicals were mixed well for 4 h using an agate mortar and were transferred carefully to alumina crucibles, and covered with a lid. Then, the powders were calcined in air at temperature 1000 °C for 10 h and then sintered out in air at 1300 °C for 16 h, with a

$$\frac{(1-x)}{2} La_2O_3 + xSrCO_3 + \frac{1}{2} Fe_2O_3 \rightarrow La_{1-x}Sr_xFeO_3 + CO_2\uparrow. \qquad \ldots\ldots\ldots (1.5.4)$$

heating rate of 5 °C/min, using a tubular electric furnace with intermediate grinding for 10 h. The sintered powders were ground again for 6 h and then compressed into circular pellets with a diameter of 12 mm and ~1mm thickness by applying a uniaxial pressure of 6 tons using a pelletizer. Finally, the pressed pellets were sintered in air at 1300 °C for 16 h. The sintered pellets of $La_{1-x}Sr_xFeO_3$ (x=0.05, 0.10, 0.15 and 0.2) multiferroics, synthesized through solid state reaction method are shown in Figure 1.5(g). Some sintered pellets of the synthesized multiferroics were finely ground again and made as powders for characterization studies.

Table 1.6 *Quantities of precursor chemicals used for the synthesis of $La_{1-x}Sr_xFeO_3$*

	Weight (gm)		
Sr concentration	**La_2O_3**	**$SrCO_3$**	**Fe_2O_3**
0.05	3.439	0.164	1.774
0.10	3.258	0.328	1.774
0.15	3.077	0.492	1.774
0.20	2.896	0.656	1.774

Figure 1.5(g) *Sintered pellets of $La_{1-x}Sr_xFeO_3$ (x=0.05, 0.10, 0.15 & 0.20)*

1.6 Characterization techniques and instrumentations

To investigate the structural, morphological, optical, magnetic and electric properties of the synthesized multiferroics the following characterization techniques were used.

i. X-ray diffraction (XRD)
ii. Scanning electron microscopy (SEM)
iii. Energy dispersive spectroscopy (EDS)
iv. UV-visible spectroscopy (UV-Vis)
v. Vibrating sample magnetometry (VSM)
vi. Dielectric measurements
vii. Ferroelectric measurements

The following section describes the principle and working of the instruments usedin the characterization techniques.

1.6.1 Powder X-ray diffraction method (PXRD)

X-ray diffraction is an accurate analytical technique that is primarily used for phase identification of crystalline solids. The experimental diffraction data can also be used to determine crystal structure through refinement. When a beam of monochromatic X-ray is focused on the sample, it interacts with the atoms of the sample and undergoes scattering. The scattered waves produce constructive interference and hence form diffraction pattern if they satisfy Bragg's law [Bragg, 1913] which is given by,

$$2d\sin\theta = n\lambda \qquad\qquad \ldots\ldots\ldots (1.6.1)$$

where, λ is the wavelength of monochromatic X-rays (1.54056 Å, for CuK_α radiation), θ is the angle of diffraction, d is the spacing between the atomic planes and n isan integer which denotes the order of the spectrum.

By measuring the angles of diffraction and intensities of the diffraction peaks, we can calculate structure factors and hence construct the three-dimensional and the two-dimensional pictures of the charge density within the crystalline solids. From this charge density analysis, the positions of the atoms within the unit cell can be determined. Consequently, the interatomic chemical bonds between the constituent atoms which are responsible for the physical properties of crystalline solids can also be investigated.

An X-ray diffractometer consists of three essential parts as an X-ray tube, a sample holder and a detector. In the X-ray tube, highly accelerated electrons are focused on the metal target (e.g., copper metal (Cu) being the most common choice). When these electrons interact with the target, characteristic X-rays with K_α and K_β components are produced. The component K_β is filtered out to produce K_α monochromatic radiation. These filtered X-rays are collimated and then directed towards the sample mounted on thegoniometer. In a powder sample, crystallites are oriented randomly so the various lattice planes are found in every possible orientation. Therefore, by scanning the sample through a particular range of 2θ angles, all possible diffractions from the lattice planes can be obtained [Azaroff, 1968]. The X-rays which satisfy Bragg's condition are diffracted by the sample and the corresponding intensity peaks are recorded by the detector. The detector processes this recorded X-rays and then converts them to count rate [Skoog et al.,2007] and gives as an output through a computer monitor or a printer. The schematic diagram of X-ray diffractometer is shown in Figure 1.6(a).

Materials Research Foundations **140** (2023)   https://doi.org/10.21741/9781644902271

In the present work, the powder X-ray diffraction characterization has been carried out at room temperature using a Bruker AXS D8 Advance X-ray diffractometer with CuKα radiation (λ=1.54056Å) at Sophisticated Analytical Instrument Facility (SAIF), STIC, Cochin University, Cochin, India.

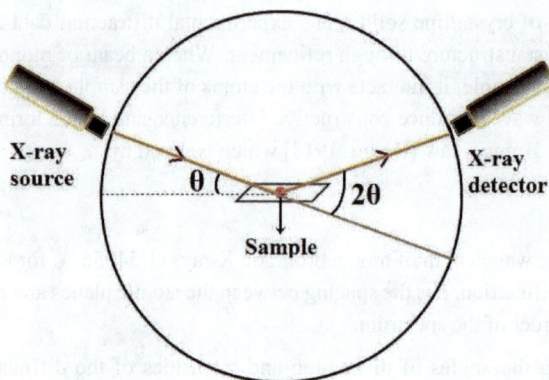

Figure 1.6(a) *Schematic diagram of X-ray diffractometer*

1.6.2 Scanning electron microscopy (SEM)

Scanning electron microscope (SEM) is one of the versatile tools for the morphological studies of the sample at a sub-micrometer length scale. SEM produces an image of the sample by scanning it with a high energy electron beam. In SEM, the electrons interact with the atoms in the sample and produce various signals. These signals contain information about the surface morphology of the sample under study. The schematic diagram of a scanning electron microscope (SEM) is shown in Figure 1.6(b). The essential components of the SEM instrument are the electron gun, lenses, sample chamber, scanning coil and detector. The electrons are ejected from the electron gun by thermionic emission and accelerated in a vacuum by the application of high electric field. These electrons are then confined into a narrow beam and directed towards the sample by a series of lenses (condenser and objective lenses). When the narrow beam of electrons hits a particular spot of the sample in the sample chamber, the surface atoms discharge secondary electrons. Most of the electrons are scattered at large angles (from 0° to 180°) when they interact with the positively charged nucleus and these scattered electrons are called back-scattered electrons (BSE). The raster scanning of the sample is achieved by changing the current in the scanning coil. From each scanning spot on the sample, secondary and back-scattered electron signals are detected by a detector and a magnified image is produced and displayed on the screen (Lawes, 1987). The image formed by secondary and back-scattered electrons contains light and dark shades. The SEM instrument requires a vacuum to prevent constant interference from the air molecules in the atmosphere.

In this work, the SEM micrographs have been recorded using the JEOL Model JSM-6390LV microscope at Sophisticated Analytical Instrument Facility (SAIF), STIC, Cochin University, Cochin, India.

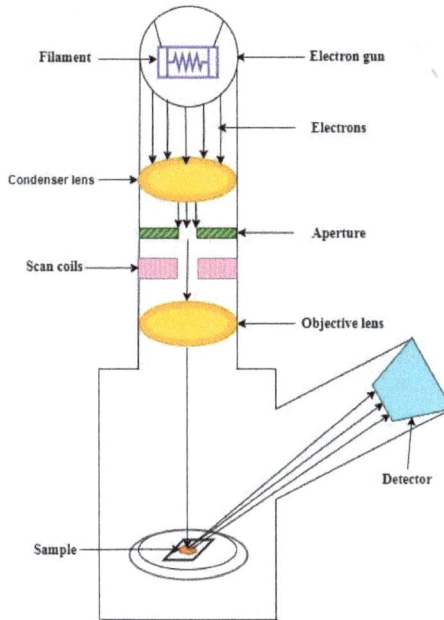

Figure 1.6(b) *Schematic diagram of scanning electron microscope*

1.6.3 Energy dispersive X-ray spectroscopy (EDS)

Energy dispersive X-ray spectroscopy (EDS) is a spectroscopic technique used for quantitative elemental analysis. Generally, the EDS system is used together with the scanning electron microscope (SEM). The principle that allows EDS to function is the capacity of high energy electron to eject innermost shell electrons from an atom in a sample.

The schematic diagram of the energy dispersive X-ray spectrometer system is shown in Figure 1.6(c). The primary components of a basic EDS system are an X-ray detector, pulse processor and analyzer. When the electron beam strikes the sample, X-rays will be released. The released X-ray beam then hits the detector which produces a charge pulse in the detector. The charge pulse is subsequently converted into a voltage pulse whose amplitude replicates the energy of the released X-ray. EDS detector is a self- contained vacuum system called a cryostat with cryogenic pumping created by liquid nitrogen cooling. FET preamplifier is used to avoid the amplification of the noise signals [Agarwal, 1991]. To avoid overlapping of two pulses, a pile-up rejector is used in EDS set up. The main amplifier gives a linear, low noise amplification of the pre-amplification signal. Finally, this analog voltage signal is converted into a digital signal by an analog to digital converter (ADC) in the computer-assisted multichannel analyzer (MCA). This digital signal causes one count to be added to the corresponding channel of MCA. The accumulated counts which are proportional to the energy of the released X-ray produce spectral peaks. EDS spectrum–a plot of X-ray counts versus energy (in keV) is displayed in digitized form. The spectral peaks are unique to the atomic structure of emitting element and hence denote a single element. Thus, EDS spectrum shows energy peaks corresponding to various elements present in the sample. Usually, these peaks are narrow and readily resolved, but many elements yield multiple peaks. The highest peak in the spectrum refers to the more concentrated element in the sample. The X-ray peaks generated by the elements in low concentration may not be resolvable from the background radiation.

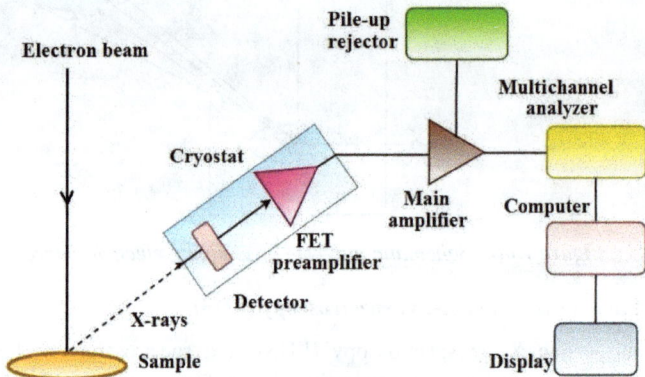

Figure 1.6(c) *Schematic diagram of an energy dispersive X-ray spectrometer (EDS) system*

In this work, the elemental compositions of the samples have been analyzed by BRUKER Nano GmbH Berlin, D-12489 at Centre for Nanoscience and Nanotechnology, The Gandhigram Rural Institute, Gandhigram, Dindigul, Tamil Nadu and JEOL Model JED-2300 at Sophisticated Analytical Instrument Facility (SAIF), STIC, Cochin University, Cochin, India.

1.6.4 Ultraviolet-visible spectroscopy (UV-Vis)

Ultraviolet-visible absorption spectroscopy concerns the absorption of near- ultraviolet (180–390 nm) or visible (390–780 nm) radiation by chemical species in the sample. This absorption causes the outer electrons to transit from the ground state to the excited energy states when the energy of the incident electromagnetic radiation matches exactly with that of energy differences between two electronic states in the atoms of the sample. It is also helpful to study the transmission, and reflectivity of various materials like coatings, pigments, etc. A UV-Vis spectrophotometer records the absorption of the material at each wavelength and presents the same in the form of a spectrum-a plot of wavelength (λ) versus absorbance (A). The optical properties of materials can be studied with the help of UV-visible absorption spectra.

The absorbance of the material is derived from Beer-Lambert's law

$$A = abc \qquad \text{......... (1.6.2)}$$

where, A is the measured absorbance (Abs), a is a wavelength-dependent absorptivity coefficient, b is the path length, and c is the analyte concentration.

The absorption coefficient (α), the incident photon energy ($h\nu$) and the optical energy gap (E_g) are related by the relation (Wood and Tauc, 1972) as

$$\alpha h\nu = K(h\nu - E_g)^n \qquad \text{......... (1.6.3)}$$

where, K is an energy independent constant and n is a constant which depends on the nature of the material and the photon transition. The values of n for direct allowed, direct forbidden, indirect allowed, and indirect forbidden transitions are 1/2, 3/2, 2 and 3 respectively. The band gap energy is determined by drawing a plot $(\alpha h\nu)^2$ versus photon energy and extrapolating the linear portion of the plot towards the zero value of $(\alpha h\nu)^2$.

A UV-visible spectrophotometer can be either a single beam or double beam. The single beam spectrophotometer was the earliest design and in this instrument, all of the light passes through the sample cell and must be measured by removing the sample. Nowadays, the double beam spectrophotometer is widely used. In a double beam arrangement, the light is separated into two beams before it reaches the sample. One beamis passed through the reference cell and another beam is passed through the sample cell. The block diagram of the double beam UV-visible spectrophotometer is presented in Figure 1.6(d). The main parts of a double beam UV-visible spectrophotometer are light source, monochromator, sample cell, reference cell and detector [Gullapalli and Barron, 2010]. The light source is normally a tungsten filament lamp for the visible region or a deuterium arc lamp for the ultraviolet region. The monochromator consists of entrance and exit slits, collimating and focusing lenses, prisms and gratings, which are used to separate different wavelengths of light. The radiation emitted from the light source is separated into its component wavelengths by the diffraction grating and dispersed by the rotating prisms. After dispersion, the spectrum is focused towards the slit at the exit. The monochromatic beam emerging out of the exit slit is then split into two beams, one passes through the reference cell and another passes through the sample cell. The intensity of the reference beam is assumed as zero absorbance; therefore, the intensity of radiation from the reference cell is stronger than that from the sample cell. Then, the two beams are detected by a detector and the ratio of these two intensity beams is measured as absorbance by the computer. Generally, photodiodes are used as detectors and they convert the light radiation into electrical signals which are read out by computer.

In this research work, UV-visible absorption spectra of the synthesized samples have been recorded using UV-visible spectrophotometer Cary 5000 (Varian Germany), in the range of 200 nm - 2000 nm wavelength at Sophisticated Analytical Instrument Facility (SAIF), STIC, Cochin University, Cochin, India.

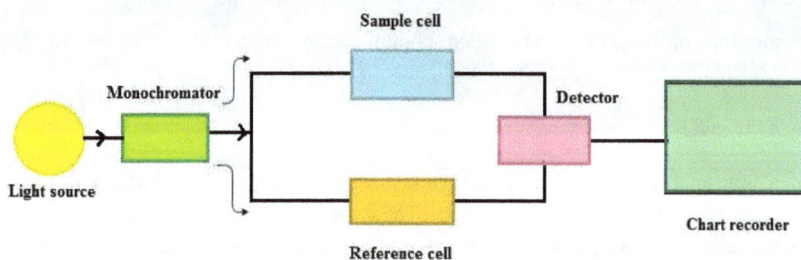

Figure 1.6(d) *Block diagram of UV-visible spectrophotometer*

1.6.5 Vibrating sample magnetometry (VSM)

The vibrating sample magnetometer is an instrument used to measure the various magnetic properties like saturation magnetization, remanent magnetization and coercive field of magnetic materials as a function of magnetic field, temperature and time. It utilizes Faraday's Law of electromagnetic induction [Foner, 1959, Wesley Burgei et al., 2003] to measure the absolute magnetic moment of a magnetic sample. Faraday's law of electromagnetic induction states that a changing magnetic field will induce an *e.m.f.*, which is given by

$$\varepsilon = -N\frac{d}{dt}(BA\cos\theta) \qquad\qquad ,\ldots\ldots\ldots (1.6.4)$$

where, ε is induced *e.m.f.* the coil, N is the number of turns in the coil, B is the applied magnetic field, A is the area of the coil, and θ is the angle between applied magnetic field B and the direction normal to the surface of the coil.

The block diagram of the vibrating sample magnetometer is presented in Figure 1.6(e). The main parts of the vibrating sample magnetometer are water-cooled electromagnet and power supply, vibration exciter and sample holder (with angle indicator), sensor coils, amplifier, control chassis, lock-in amplifier and computer Interface [Foner, 1959]. The magnetic material under study is placed on the sample holderrod and is centered in a pair of pick up coils between the poles of an electromagnet. The electromagnet generates the constant magnetic field which is used to magnetize thesample. The sample holder rod is attached with the vibration exciter which moves the sample up and down at a set frequency usually 85 Hz. Control chassis controls the 85 Hz oscillation of the exciter. By rotating the sample rod, the desired orientation of the sample is set at a constant magnetic field. The vibrating sample now produces an alternating current in sensor (pickup) coils mounted on the pole pieces of the magnet. This induced AC signal which is at the vibration frequency is proportional to the magnetic moment of the sample, vibration amplitude and frequency. The signal produced in the sensor coil is amplified by an amplifier. The lock-in amplifier is adjusted to pick up only signals at the vibrating frequency by eliminating noise from the environment. The computer Interface isutilized for the data collection and the resultant data can be graphed and plotted on the printer.

In this work, the magnetic measurements have been carried out using Lakeshore VSM 7410 model at SAIF, IIT Madras, Chennai, India.

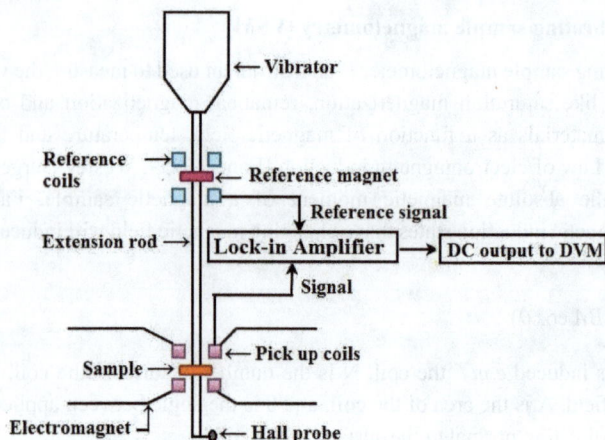

Figure 1.6(e) Block diagram of Vibrating sample magnetometer (VSM)

1.6.6 Dielectric measurements

The study of dielectric measurements helps to understand the behavior and mechanism of the electrical conduction through polarization in the sample. Dielectrics are generally characterized by dielectric constant, dielectric loss and dielectric conductivity. The dielectric constant (ε') is defined as the ratio of the absolute permittivity of the material to the permittivity of free space. Dielectric loss is the amount of electrical energy which is dissipated through conduction when an alternating electric field is applied across the dielectric material. It can be measured in terms of either the loss angle δ or the corresponding loss tangent tan δ. The ac conductivity is a measure of electrical conduction that arises due to the hopping of charge carriers on the application of an electric field. The dielectric properties are measured as a function of frequencies at room temperature using an Impedance analyzer. The analyzer is interfaced with the computer, and the data such as capacitance, dissipation factor and inductance are collected as a function of frequencies. The schematic diagram of the circuit used in the dielectric measurements set up is presented in Figure 1.6(f). For dielectric measurements, the pellet of the ceramic sample is polished and coated with the silver paste in order to make good electrical contact and to form parallel plate capacitor (C_S) geometry. This sample is mounted between two electrodes forming a parallel plate capacitor. Then this capacitor (C_S) is connected in series with a reference capacitor (C_r) and is subjected to an AC voltage using an AC signal generator. The voltage across the reference capacitor (V_r) and the sample capacitor (V_s) is measured by two separate voltmeters. Since the two capacitors are connected in series, the charge on the reference capacitor and the sample capacitor (Q_s) will be the same. So, the charge on the sample is found by:

$$Q_s = Q_r = C_r \times V_r \qquad \text{......... (1.6.5)}$$

Then by using the measured voltage across the sample the capacitance of the sample is determined by the following equation:

$$C_s = \frac{Q_s}{V_s} \qquad \text{......... (1.6.6)}$$

The dielectric constant s' is determined using the formula:

$$s' = \left| \frac{(C_s \times t)}{\varepsilon_0\, A} \right| \qquad \text{......... (1.6.7)}$$

where, C_s is the capacitance of the sample, t is the thickness of the pellet, ε_0 is the permittivity of free space ($\varepsilon_0 = 8.854 \times 10^{-12}$ F/m) and A is the area of the pellet.

Generally, the capacitance of a parallel plate capacitor (with and without dielectric material) having two metal electrodes of area A, separated by a distance d is represented by the following expressions,

$$C = \frac{s A}{d} \qquad \text{......... (1.6.8)}$$

$$C_0 = \frac{s_0 A}{d} \qquad \text{......... (1.6.9)}$$

where, C and C_0 represent the capacitance of the capacitor with and without dielectric material respectively, ε and ε_0 represent the permittivity of the material and permittivity of free space ($\varepsilon_0 = 8.854 \times 10^{-12}$ F/m) respectively. Then, the measured capacitances are used to find the relative permittivity using the following formula:

$$\varepsilon = \frac{C}{C_0} = \frac{s}{s_0} \qquad \text{......... (1.6.10)}$$

During dielectric measurements, the electrical energy is absorbed by the material which is dissipated in the form of heat and the dissipation is called dielectric loss (tan δ). The dielectric loss is related to the imaginary part of the permittivity, and it is often described by the tangent of the dielectric loss (tan δ).

$$\tan \delta = \frac{\varepsilon''}{\varepsilon'} \qquad \dots\dots\dots (1.6.11)$$

where, δ is the angle between the imaginary and real components of the permittivity.

In this research work, the frequency-dependent dielectric measurements have been carried out at room temperature using an impedance analyzer PSM 1735 N4L, at Abraham Panampara Research Center (APRC), Sacred Heart College, Vellore, Tamilnadu, India, as shown in Figure 1.6(g).

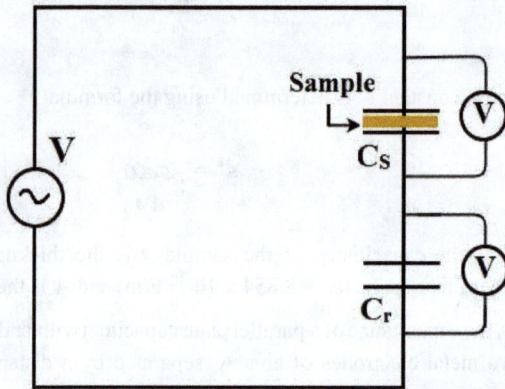

Figure 1.6(f) *The schematic diagram of the circuit used in the dielectric permittivity measurements*

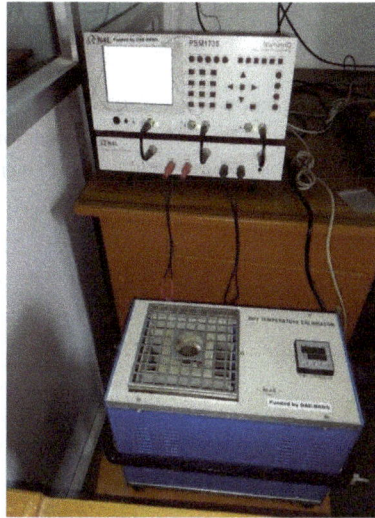

Figure 1.6(g) *Photograph of impedance analyzer PSM 1735*

1.6.7 Ferroelectric measurements

The ferroelectric properties of a material are analyzed by tracing a ferroelectric hysteresis loop using a P-E Loop Tracer. The ferroelectric hysteresis loop is a plot of the charge or polarization (P) developed in the material on application of an external electric field (E) at a given frequency. The electric polarization measurement method is based on Sawer-Tower circuit [Sawer and Tower, 1930] and the schematic diagram of the circuit is shown in Figure 1.6(h). It consists of two capacitors, C_s (s stands for the ferroelectric sample) and C_r in series (r stands for reference) and load resistance R, which is a series combination of R_1 and R_2. By choosing an appropriate value for C_r and R, a step voltage

(V) is applied by a signal generator on the surface of electrodes of a sample capacitor (C_s). The voltage (V_r) across the reference capacitor (C_r) is measured. As the capacitors are connected in series, the charge ($Q_r = C_r \times V_r$) on the reference capacitor must be the same as that over the sample capacitor. Once the charge (Q_s) on the ferroelectric sample is known by the measurement the electric polarization (P) can be calculated using the expression,

$$P = \frac{Q_s}{}$$

$$P = \frac{Q_s}{A}$$

.... (1.6.12)

where, Q_s is the charge developed on the electrodes of the ferroelectric sample capacitor and A is the area of the electrodes. In an electric polarization measurement, the opposite surfaces of the samples are polished and coated with silver paste in order to make the electrodes and hence to form parallel plate capacitor geometry.

In this research work, the ferroelectric measurements have been done using a Ferroelectric loop tracer, Marine India Pvt. Ltd, at B. S. Abdur Rahman Crescent Institute of Science and Technology, Chennai, India.

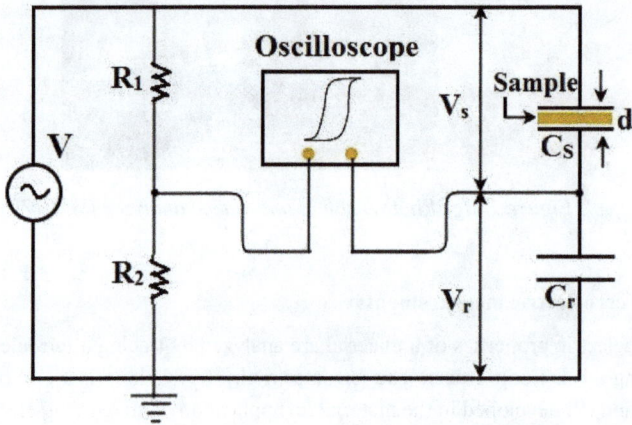

Figure 1.6(h) *Schematic circuit of the Sawyer-Tower Bridge for measuring the P-E characteristics of ferroelectrics*

1.7 Methodologies employed for analysis

In this research work, the synthesized samples were subjected to various analytical methodologies, in addition to the experimental characterizations discussed in the previous section. This section deals with the methodologies namely Rietveld [Rietveld, 1969] technique and Maximum Entropy Method (MEM) [Collins, 1982] employed to refine the raw X-ray diffraction data and to analyze charge density distribution. This section also deals with the Tauc plot methodology [Wood and Tauc, 1972] and Scherrer [Cullity, 2001] formula employed for the calculation of energy band gap and grain size.

1.7.1 XRD powder profile refinement through Rietveld technique

Rietveld refinement [Rietveld, 1969] is a powder profile refinement technique for use in the structure determination of crystalline materials. This method was devised by Hugo Rietveld. In the polycrystalline sample, due to the random orientation of crystallites the XRD peaks of the powder diffraction pattern overlap which may lead to some loss of information about the synthesized sample and prevent the exact determination of the crystal structure. The Rietveld refinement [Rietveld, 1969] techniquereduces the effect of these overlapping peaks by calculating the expected intensity for every individual step in the X-ray pattern during structural refinement and hence determines the real crystal structure. In Rietveld method [Rietveld, 1969] least squares approach is employed to refine a theoretically constructed profile until it matches the experimentally observed profile and the profiles are refined by iterative technique. At present, Rietveld method [Rietveld, 1969] has been so successful in determining the structure of polycrystalline materials which are in the form of powders and so it is regarded as an elemental technique of characterizing polycrystalline materials in the field of physics, chemistry, materials science, mineralogy, etc.

1.7.1.1 Rietveld refinement procedure

Rietveld refinement [Rietveld, 1969] method requires high quality powder diffraction data of the material in step scan mode possibly with $0.02°$ increment in 2θ range from $10°$ to $120°$. The principle of the Rietveld refinement [Rietveld, 1969] method is to minimize a function M which scans the weighted sum of the squared differences between the experimentally observed data y(obs) and a calculated profile y(cal). The function M is defined as,

$$M = \sum_i W_i \left\{ y_i^{obs} - \frac{1}{c} y_i^{calc} \right\}^2 \qquad \ldots\ldots\ldots (1.7.1)$$

where, $\sum$ is the sum of independent observations, W_i is the statistical weight and C is the overall scale factor.

The experimental powder X-ray diffraction pattern contains intensity distributionwith respect to the Bragg angle. The calculated intensity y^{cal} at a given point of the diffractogram is obtained using the formula

$$y^{cal} = S \, \Sigma_k L_k \, |F_k|^2 \, \phi \, (2\theta_i - 2\theta_k) P_k A + y_{bi} \qquad \ldots\ldots\ldots (1.7.2)$$

where, S is the scale factor, k is the Miller indices (*hkl*) of a Bragg reflection, L_k is the Lorentz polarization and multiplicity factors, F_k is the structure factors, $\phi\,(2\theta_i - 2\theta_k)$ is the reflection profile function, P_k is the preferred orientation function, A is the absorption factor and y_{bi} is the background intensity.

In this work, the software program JANA 2006 [Petricek et al., 2014] has been employed for the XRD profile refinement. In Rietveld method [Rietveld, 1969] the structural parameters such as lattice parameters, atomic fractional coordinates, occupancies, isotropic/anisotropic atomic displacement parameters, etc., are refined in addition to the scale factor, preferred-orientation, peak shape functions such as Gaussian, Lorentzian and Psedo-voight. The number of refinable parameters increases significantly when dealing with samples consisting of two or more phases. The quality of the Rietveld [Rietveld, 1969] refinement between the theoretically constructed and observed profiles is indicated by some residual functions. The profile R-factor is a measure of the agreement between the observed and calculated profile. The quality of the profile fit can be monitored by the profile factor (R_P), the weighted profile factor (R_{wp}) and the goodness of fit (χ^2) on each step of the iteration. The profile fit is best, when the value of $\chi^2 = 1$, indicating the perfect match between experimental and theoretical data. The R-factors are defined by the following equations:

$$R_p = \frac{\sum ||F_{obs}| - |F_{cal}||}{\sum |F_{obs}|} \qquad \cdots\cdots (1.7.3)$$

$$R_{wp} = \frac{\sum w_i ||F_{obs}| - |F_{cal}||^2}{\sum w_i |F_{obs}|^2} \qquad \cdots\cdots (1.7.4)$$

$$\gamma^2 = \frac{\sum w_i ||F_{obs}| - |F_{cal}||^2}{N - P} \qquad \cdots\cdots (1.7.5)$$

Where F_{obs} and F_{cal} are the observed and calculated structure factors, N is the number of data points and P is the number of refined parameters.

1.7.1.2 Peak shape

In powder X-ray diffraction, the shape of a peak is influenced by the characteristics of the incident X-ray beam, the sample size and shape and the experimental arrangements. The experimental profile of a single diffraction peak is empirically assumed as Gaussian in shape. If this type of distribution is assumed, then the contribution of a given reflection to the experimental profile y_i at the position $2\theta_i$ is :

$$y_i = I_k \exp\left[\frac{-4\ln(2)}{H_k^2}(2\theta_i - 2\theta_k)^2\right] \qquad \cdots\cdots (1.7.6)$$

where, I_k is the calculated intensity of the reflection, H_k is the full w

idth at half maximum (FWHM), $2\theta_k$ is the position of the peak. Due to the use of the finite slit heights, together with finite sample heights, the diffraction peaks may acquire an asymmetry at very low diffraction angles. The vertical divergence effect [Klug and Alexander, 1959] causes the shift of the maximum of the diffraction peak to lower angles without disturbing the integrated peak area. To account for this asymmetry Rietveld introduced a semi-empirical correction factor

$$A_s = 1 - \left[\frac{P(2\theta_i - 2\theta_k)^2}{\tan\theta_k} \right]$$

$$\ldots\ldots (1.7.7)$$

where, P is the asymmetry factor and s takes different values as, +1, 0, -1 depending on the difference $(2\theta_i-2\theta_k)$ being +ve, 0, −ve respectively. Actually, at a given position, more than one diffraction peak may contribute to the profile. Therefore, the intensity of the diffraction peak is simply the sum of all the reflections contributing at the given point $2\theta_i$.

1.7.1.3 Peak width

In powder X-ray diffraction pattern, the width of the XRD peaks is found to broaden at higher angles due to the particle size effect. The angular dependency is expressed by the formula [Caglioti et al., 1958],

$$H_k^2 = U\tan^2\theta_k + V\tan^2\theta_k + W \qquad \ldots\ldots (1.7.8)$$

where, U, V & W are half-width parameters and may be refined during the fit.

1.7.1.4 Preferred orientation

In powder crystal samples, at least in part of the samples, the crystallites will be rod-like or plate-like and they tend to align themselves along the axis of a cylindrical sample holder. In solid polycrystalline samples, the synthesis of the material may result in a greater volume fraction of certain crystal orientations. In such situations, the reflection

intensities will vary from those predicted for a completely random distribution. The corrected factor used by Rietveld [Rietveld, 1969] to obtain the corrected intensity for the preferred orientation is given as,

$$I_{corr} = I_{obs} \, exp(-G\alpha^2) \qquad \ldots\ldots (1.7.9)$$

where, I_{obs} denotes the observed intensity for a random sample, G denotes the preferred orientation parameter and α is the acute angle between the scattering vector and the normal of the crystallites.

1.7.1.5 Background function

The background, y_{bi}, at step i, which is approximated by a finite sum of Legendre polynomials, $F_j(x_i)$ [Abramowitz and Stegun, 1964], orthogonal relative to integration over

$$y_{bi} = \sum_{j=0}^{\Pi} b_j F_j(x_i)$$

     (1.7.10)

the interval [-1, 1] is,

$F_j(x_i)$'s for j 2 are computed from $F_{j-1}(x_i)$ and $F_{j-2}(x_i)$ using the relation,

$$F_j(x_i) = \left(\frac{2j\pm1}{j}\right) x_i \ F_{j\pm1}(x_i) \pm \left(\frac{j\pm1}{j}\right) F_{j\pm2}(x_i)$$

     (1.7.11)

$$x_i = \frac{2\theta_i - \theta_{max} - \theta_{min}}{\theta_{max} - \theta_{min}}$$

     (1.7.12)

with the values $F_0(x_i) = 1$ and $F_1(x_i) = x_i$. The coefficients, b_j, are called background parameters to be refined in Rietveld [Rietveld, 1969] technique, and the variable, x_i is normalized between -1 and 1 as follows,

The correlation coefficients between background parameters can be brought down to some extent by means of this background function. Furthermore, the "humps" due to amorphous or poorly crystallized material may also be fitted well by doing the refinement with more refinable background parameters.

1.7.1.6 Rietveld refinement using JANA 2006

In this work, the software program JANA 2006 [Petricek et al., 20 14] has been employed to perform Rietveld [Rietveld, 1969] refinement for all the synthesized multiferroics. JANA 2006 [Petricek et al., 2014] is a crystallographic software program for structural refinement of composite as well as aperiodic crystals based on the X-ray diffraction patterns. It can refine the crystal structure having two or more phases. During refinement, the experimentally observed XRD profiles are matched with the theoretically constructed profiles using pseudo-Voigt [Wertheim, 1974], profile shape functions [Thompson, 1987] and Gaussian FWHM parameters.

The profile symmetry is also included by using Simpson rule of integration given by Howard [Howard, 1982], which incorporates symmetric profile shape function with various coefficients and peak shift. JANA 2006 [Petricek et al., 2014] also uses the corrections for preferred orientation using March-Dollase function [March, 1932, Dollase, 1986]. The theoretical profiles thus constructed are compared with the observed ones. Finally, the structure factors evolved from the Rietveld [Rietveld, 1969] refinement were further utilized for analyzing charge density distribution in the unit cell.

1.7.2   Charge density analysis through Maximum Entropy Method (MEM)

To analyze the electronic properties of the materials, the electronic charge distribution and bonding nature between the constituent atoms of the materials are needed. Therefore, the ultimate aim of this work is to thoroughly analyze the precise electronic structure of the lanthanum orthoferrite-type multiferroics and the bonding interaction between the constituent atoms which are responsible for the physical properties of the synthesized multiferroics. This aim is achieved successfully by adapting Maximum Entropy Method (MEM) [Collins, 1982]. The following sections, describe the importance of electron density studies in structure analysis, the formalism of MEM method [Collins, 1982], the principle, MEM [Collins, 1982] methodology followed in this work to elucidate the electronic charge density distribution in the unit cell of the synthesized multiferroics.

1.7.2.1 Electron density

The electron density is defined as the number of electrons per unit volume. In general, the regions of electron density which are called electron clouds are found around the atoms. The quantum mechanical theory tells us that electron density is the measure of the probability of an electron being present at a particular location. The electron density distribution in the unit cell of the crystalline samples can be estimated through scattering experiments, especially, X-ray diffraction from the crystals. In diffraction experiments, the relative intensities of the reflections are measured in order to deduce the electron density distribution in the unit cell. The electron density distribution is associated with the diffraction pattern intensities. The diffraction intensities are the Fourier transforms of the crystal structure and in turn, the crystal structure is the Fourier transform of the diffraction intensities which is expressed in respect of electron density distribution concentrated in atoms. This cannot be determined by direct experimental methods, as the scattered X-rays cannot be refracted by lenses to form an image as done with light in an optical microscope.

Materials Research Forum LLC

https://doi.org/10.21741/9781644902271

Furthermore, it cannot be obtained directly, by calculation as the relative phases of the waves are unknown. Therefore, the electron density distribution can be determined using the Fourier series if it is given a set of structure factors.

The probability of locating an electron at one point or another can be estimated quantum mechanically and this estimation gives a quantity called electron density. Electron density has all the information like the interatomic chemical bonding, physical and chemical properties of the crystal systems. The electron density distribution study is used in basic sciences like physics, chemistry, biology and geology (Stout and Jenson, 1968). The study of electronic structure and interatomic chemical bonding of crystalline materials is very important and it provides useful information about the transport properties which can be effectively utilized for device applications.

$$\rho(r) = \frac{1}{V}\sum_H F(H)\exp(-2\pi iH.r) \qquad \ldots\ldots\ldots (1.7.13)$$

The electron density $\rho(r)$ is derived from the structure factor as

Where V is the unit cell volume, $F(H)$ are structure factors and H are indices denoting a particular scattering direction corresponding to a crystal plane.

Since a crystal lattice has a periodicity, the electron density in a crystal can be described as a periodic function. The number of electrons in any volume element dV is given by $\rho(x, y, z)dV$. In an X-ray diffraction experiment, when the X-rays are used to scan the sample, the wavelet scattered by this volume element will be,

$$\rho(x, y, z)\exp[-2\pi i(hx + ky + lz)]dV \qquad \ldots\ldots\ldots (1.7.14)$$

If we consider the whole unit cell, the resultant sum of contributions from all such volume elements in the unit cell will be,

$$F_{hkl} = \int \rho(x, y, z)\exp[-2\pi i(hx + ky + lz)]dV \qquad \ldots\ldots\ldots (1.7.15)$$

where, F_{hkl} is the structure factor. Thus, the structure factor is the result of the addition of the waves scattered in the direction of the (hkl) reflection from the atoms in the unit cell. If the structure factors and phases are known, the electronic charge density distribution of the unit cell can be determined. There must be a one to one relationship between observed diffraction intensities and electron density i.e., a given set of intensities must correspond to one and only one electron density distribution. The magnitudes of individual structure factors are determined as the square root of the experimentally observed diffraction intensity and their phases are determined by solving the structure. The interpretation is taken as a model and is improved by least-squares refinement based on the structure factors to obtain an accurate model. The electron density in the unit cell can then be obtained by taking the Fourier summation of phased structure factors.

1.7.2.2 Electron density derived from Fourier method

A series of trigonometric terms that are necessary to describe a well-behaved continuous periodic function is called a Fourier series. It is assumed that the unit cell is partitioned into small volumes dV in which there are $\rho(r)dV$ number of electrons. The amplitude of the wave scattered from such a small volume element will be $\rho(r)dV$ times the amplitude of wave scattered from an electron at the same position. Therefore, the total scattered amplitude can be estimated from the electron density distribution $\rho(r)$. $F(H)$ can be written in terms of electron density $\rho(r)$ as

$$F(H) = \int \rho(r)\, exp(2\pi i H. r)\, dV \qquad \ldots\ldots\ldots (1.7.16)$$

The inverse Fourier transform of the above expression gives the electron density as,

$$\rho(r) = \int F(H)\, exp(-2\pi i H. r)\, dV = \frac{1}{V} \sum F(H) exp(-2\pi i H. r) \qquad \ldots\ldots\ldots (1.7.17)$$

Since F(H) is defined as the discrete set of reciprocal lattice points k, the integral is replaced by the summation. The structure factor can be written as,

$F(H) = A(H) + iB(H)$, then,

$$\rho(r) = \frac{1}{V}\sum (A(H) + iB(H))[cos(2\pi H.r) - i\, sin(2\pi H.r)] \qquad \ldots\ldots\ldots (1.7.18)$$

Where, $A(H)=A(-H)$ and $B(H)=B(-H)$, since the electron density function is real.

Therefore, $\rho(r) = \frac{1}{V}\sum_{1/2} [2A(H)\, cos(2\pi H.r) + 2B\, sin(2\pi H.r)] \qquad \ldots\ldots\ldots (1.7.19)$

Taking $A(H) = |F(H)|\, cos\, \varphi$ and $B(H) = |F(H)|\, sin\, \varphi$ the electron density becomes,

$$\frac{\rho(r)}{V} = 2 \sum_{1/2} [|F(H)|\, cos\, \varphi\, cos(2\pi H. r) + |F(H)|\, sin\, \varphi\, sin(2\pi H. r)] \ldots (1.7.20)$$

which further becomes,

$$\rho(r) = \frac{2}{V}\sum_{1/2}[|F(H)|\, cos(2\pi H.r) - \varphi(H)] \qquad \ldots\ldots\ldots (1.7.21)$$

From the expression (1.7.21) for electron density, it is obvious that each structure factor contributes a plane wave to the total electron density with wave vector H and phase φ. Thus, to estimate the electron density, phases of the structure factors need to be known.

If we know the approximation to the scattering density, then the phase $\varphi(H)$ can be found and an imperfect image of the structure can be derived. Anomalous scattering should also be corrected by subtracting the calculated $\Delta A_{calc}^{anamalous}$ and $\Delta B_{calc}^{anamalous}$ contributions

from A and B, respectively, using the anomalous scattering factors f' and f''.

The period of the plane wave with the amplitude $F(H)$, in the direction of the wave vector H is $1/H$. So, the period is shorter for higher-order reflections which include resolution to the image. As more higher-order reflections are included, a higher resolution of image can be obtained, which is similar to the enhancement of resolution in an optical image obtained with shorter wavelength radiation.

To obtain the Fourier transform of the diffraction pattern as the image, it is necessary to adapt computational methods, since the lenses cannot be used for X-ray beams. To find the charge density using the Fourier method, an infinite number of Fouriercoefficients are needed. But in practice, only a limited number of Fourier coefficients are used, by ignoring the experimental errors and setting all the missing Fourier coefficients as zero. This assumption is a highly biased one and it leads to the negative electron density. So, this will not provide accurate charge density properties. Therefore, in the present work, precise charge density studies have been performed by adapting the Maximum Entropy Method which is called as MEM [Collins, 1982].

1.7.2.3 MEM methodology

The charge or electron density distribution and interatomic bonding analysis of a crystal structure are useful in correlating the physical properties and atomic arrangement of material. In this work, the Maximum Entropy Method is employed, to analyze the spatial electron density distribution and interatomic chemical bonding inside the unit cell of the synthesized multiferroic materials and to correlate their physical properties. In MEM, the structure factors obtained from the Rietveld [Rietveld, 1969] refinement are utilized. In the MEM refinement process, all the structure factors are refined using Fortran 90 software program PRIMA (PRactice Iterative MEM analysis) [Izumi and Dilanien, 2002] to construct the charge density distributions in the unit cell. PRIMA [Izumi and Dilanien, 2002] is a software program to evaluate charge densities from experimental X-ray diffraction data and its input file contains the cell parameters, space group, Lagrange parameter, pixels, total number of electrons in the unit cell and structure factors.

In this work, the unit cell of the synthesized samples was partitioned into 48 x 72 x 48 pixels along the a, b and c axes of the orthorhombic lattice and the initial charge density in each of the pixel was uniformly fixed as Z/a_0^3, where Z is the total number of electrons in the unit cell and a_0^3 is the volume of the unit cell. During MEM charge density computations, the Lagrangian multiplier is selected suitably in each case such that, the convergence criterion C becomes unity after executing the minimum number of iterations. To visualize electron density distribution in the unit cell clearly, the three-dimensional (3D) electron density was plotted using VESTA (Visualization for Electron and STructural Analysis) [Momma and Izumi, 2008] software package. VESTA [Momma and Izumi, 2008] represents the crystal structure by using different models like ball and stick, space filling, polyhedral and wireframe along with volumetric data at the same window. To realize the interaction and nature of chemical bonding between the atoms in the synthesized multiferroics, the two-dimensional (2D) and one-dimensional (1D) charge densities on different lattice planes have been mapped and explained in the following chapters.

1.7.3   Energy band gap determination using UV-visible absorption spectra

The energy band gap of a material can be determined from the UV-visible absorption/reflection spectrum. In this work, UV-visible absorption data of the samples have been used for the determination of the energy band gap. The energy band gap (E_g) is related with the absorbance (α) and photon energy (hv) by the following Woodand Tauc [Wood and Tauc, 1972] relation,

$$\alpha h\upsilon = A(h\upsilon - E_g)^n \qquad\qquad \text{......... (1.7.22)}$$

where, A is an energy independent constant and n is a constant which depends on the nature of the material (crystalline or amorphous) and the photon transition. The constant n=1/2 for direct allowed transition, n=2 for indirect allowed transition, n=3/2 for direct forbidden transition and n=3 for indirect forbidden transition. For direct band gap materials n=1/2. Hence, equation (1.7.23) can be written as,

$$\alpha h\upsilon = A(h\upsilon - E_g)^{1/2} \qquad\qquad \text{......... (1.7.23)}$$

By squaring equation (1.7.23),

$$(\alpha h\upsilon)^2 = A(h\upsilon - E_g) \qquad\qquad \text{......... (1.7.24)}$$

The above equation resembles the equation of a straight line,

$$y = mx + C \qquad\qquad \text{......... (1.7.25)}$$

Comparing the two equations (1.7.24) and (1.7.25),

$$y = (\alpha h v)^2 \qquad \dots\dots (1.7.26)$$

If y=0, then the equation (1.7.25) becomes,

$$A(hv - E_g) = 0 \qquad \dots\dots (1.7.27)$$

In equation (1.7.41), $A \neq 0$,

Therefore $(hv - E_g) = 0$

Hence,

$$E_g = hv \qquad \dots\dots (1.7.28)$$

Equation (1.7.42) gives the energy band gap, which was evaluated using Tauc's procedure. Thus in this work, to determine the energy band gap of the synthesized multiferroic materials using UV-visible absorption data, a Tauc plot has been drawn by taking the photon energy (hv) in the x-axis, and $(\alpha h v)^2$ in the y-axis. Then, by extrapolating the tangent of the linear portion of the Tauc plot to meet the x-axis where $(\alpha h v)^2 = 0$, the energy band gap E_g is determined.

1.7.4 Grain size evaluation

In this work, the average grain or crystallite size was evaluated from the experimental powder X-ray diffraction data using the Scherrer formula [Cullity and Stock, 2001] through GRAIN software program [Saravanan, 2008]. The Scherrer formula [Cullity and Stock, 2001] is given as,

$$t = \frac{0.9\lambda}{\beta \cos\theta} \qquad \dots\dots (1.7.29)$$

where, t is the average grain or crystallite size (size of coherently diffracting domains), λ is the wavelength of X-ray used, which is 1.54056 Å, β is the Full width at half maximum (FWHM) in radians, and θ is the Bragg angle.

For the GRAIN software program [Saravanan, 2008], the Bragg angles and their corresponding full-width at half maxima (FWHM) obtained from the powder X-ray diffraction data were given as input data.

48

References

[1] Abramowitz M., Stegun I.A., Handbook of Mathematical Functions, National Bureau of Standards, (1964).

[2] Acharya S., Chakrabarti, P.K., solid state commun., 150, 1234 (2010) . https://doi.org/10.1016/j.ssc.2010.04.006

[3] Acharya S., Mondal J., Ghosh S., Roy S.K., Chakrabarti P.K., Mater. Lett., 64, 415 (2010). https://doi.org/10.1016/j.matlet.2009.11.037

[4] Adler D., Feinleib J., Phys. Rev. B, 2, 3112 (1970). https://doi.org/10.1103/PhysRevB.2.3112

[5] Agarwal B.K., X-ray spectroscopy, 2nd edition, Springer-verlog, Berlin, (1991). https://doi.org/10.1007/978-3-540-38668-1

[6] Ahmed M.A, Azab A.A., El−Khawas E.H., J. Mater. Sci: Mater. Electron., 26, 8765 (2015). https://doi.org/10.1007/s10854-015-3556-4

[7] Anderson P.W., Phys. Rev., 79, 350 (1950). https://doi.org/10.1103/PhysRev.79.350

[8] Azaroff L.V., Elements of X-ray crystallography, Mc Graw hill book company, New York, 79 (1968).

[9] Bhat I., Husain S., Khan W., Patil S.I., Mater. Res. Bull., 48, 4506 (2013). https://doi.org/10.1016/j.materresbull.2013.07.028

[10] Birss R.R., Symmetry and magnetism. Vol. 3, North-Holland publ. Co., Amsterdam, pp.128-129, 133-136, (1964).

[11] Bragg W.L., The diffraction of short electromagnetic waves by a crystal. Proc Cambridge Philoso Soc. 17 (1913).

[12] Brink J.V.D., Khomskii D.I., J. Phys.: Condens. Matter., 20, 434217 (2008). https://doi.org/10.1088/0953-8984/20/43/434217

[13] Caglioti G., Paoletti A., Ricci F.P., Nucl. Instrum., 3, 223 (1958). https://doi.org/10.1016/0369-643X(58)90029-X

[14] Cheong S.W., Mostovoy M., Nat. Mater., 6, 13 (2007). https://doi.org/10.1038/nmat1804

[15] Chougule S.S., Chougule B.K., Smart Mater. Struct., 16, 493 (2007). https://doi.org/10.1088/0964-1726/16/2/030

[16] Collins D.M., Nature, 298, 49 (1982). https://doi.org/10.1038/298049a0

[17] Cullity B.D., Stock S.R., Elements of X-ray diffraction, Pearson education. 3rd edn.

Prentice Hall, Upper Saddle River, 558 (2001).

[18] Das S.R., Choudhary R.N.P., Bhattacharya P., Katiyar R.S., Dutta P., Manivannan A., Seehra M.S., J. Appl. Phys., 101, 034104 (2007). https://doi.org/10.1063/1.2432869

[19] Devi P.S., Sharma A.D., Maiti H.S., Indian Ceram Soc., 63, 75 (2004). https://doi.org/10.1080/0371750X.2004.11012140

[20] Dollase W.A., J .Appl. Crystallogr., 19, 267 (1986). https://doi.org/10.1107/S0021889886089458

[21] Dong S., Liu J.M., Cheong S.W., Ren Z.F., Adv. Phys., 64, 519 (2015). https://doi.org/10.1080/00018732.2015.1114338

[22] Eerenstein W, Mathur N. D, Scott J. F, Nature, 442, 759 (2006). https://doi.org/10.1038/nature05023

[23] Fiebig M., Lottermoser T., Meier D., Trassin M., Nat. Rev. Mater., 1, 16046 (2016). https://doi.org/10.1038/natrevmats.2016.46

[24] Foner S., Rev. Sci. Instrum., 30, 7 (1959) https://doi.org/10.1063/1.1716679

[25] Gajek M., Bibes M., Fusil S., Bouzehouane K., Fontcuberta J., Barthélémy A., Fert A., Nat Mater., 6, 296 (2007). https://doi.org/10.1038/nmat1860

[26] Giannakas A.E., Leontiou A.A , Ladavos A.K., Pomonis P., J, Appl. Catal. A, 309, 254 (2006). https://doi.org/10.1016/j.apcata.2006.05.016

[27] Goldschmidt V.M., Die Gesetze der Krystallochemie, DieNaturwissenschaften, 14, 477 (1926). https://doi.org/10.1007/BF01507527

[28] Gullapalli S., Barron A.R. , 'Characterization of Group 12-16 (II-VI) Semiconductor Nanoparticles by UV-visible Spectroscopy', OpenStax CNX, June, 2010 Online: Web site. http://cnx.org/content/m34601/1.1/, June 12, 2010.

[29] Hearne G.R., Pasternak M.P., Phys. Rev. B: Condens. Matter., 51, 11495 (1995). https://doi.org/10.1103/PhysRevB.51.11495

[30] Hill N.A., J. Phys.Chem.B., 104, 6694 (2000). https://doi.org/10.1021/jp000114x

[31] Howard C.J., J. Appl. Crystallogr., 15, 615 (1982). https://doi.org/10.1107/S0021889882012783

[32] Hur N., Park S., Sharma P.A., Ahn J.S., Guha S., Cheong S.W., Nature, 429, 392 (2004). https://doi.org/10.1038/nature02572

[33] Izumi F., Dilanien R.A., Recent Research Developments in Physics Part II, Vol.3,

Transworld Research Network. Trivandrum, p699, (2002).

[34] Janbutrach Y., Hunpratub S., Swatsitang E., Nanoscale. Res. Lett., 9, 498 (2014). https://doi.org/10.1186/1556-276X-9-498

[35] Jaynes E.T., IEEE Trans Syst Sci Cybern SSC. 4, 227 (1968). https://doi.org/10.1109/TSSC.1968.300117

[36] Kharton V.V., Solid State Electrochemistry II: Electrodes, Interfaces and Ceramic Membranes, 1st edition, Wiley-VCH, Weinheim (2011). https://doi.org/10.1002/9783527635566

[37] Khomskii D., Physics, 2, 20 (2009). https://doi.org/10.1103/Physics.2.20

[38] Kimura T., Goto T., Shintani H., Ishizaka K., Arima T., Tokura Y., Nature, 426, 55 (2003). https://doi.org/10.1038/nature02018

[39] Klug H.P., Alexander L.E., X-ray diffraction procedures, 2nd edition, John Wiley New York, 251 (1959).

[40] Koehler W.C., Wollan E.O., Wilkinson M.K., Phys. Rev., 1, 58 (1960). https://doi.org/10.1103/PhysRev.118.58

[41] Koops C.G., Phys. Rev., 83, 121(1951). https://doi.org/10.1103/PhysRev.83.121

[42] Kozhevnikov V.L., Leonidov I.A., Patrakeev M.V., Markov A.A., Blinovskov Y. N., J Solid State Electrochem., 13, 391 (2009). https://doi.org/10.1007/s10008-008-0572-9

[43] Landau L.D., Lifshitz E.M., Pitaevskii L.P., Electrodynamics of continuous media. 2nd edition, Elsevier, Heidelberg (1984). https://doi.org/10.1016/B978-0-08-030275-1.50007-2

[44] Lawes G., Scanning electron microscopy and X-ray microanalysis: Analytical chemistry by open learning, John Wiley & sons, (1987).

[45] Lee N., Choi Y.J., Ramazanoglu M., Ratcliff W., Kiryukhin V., Cheong S.W., Phys. Rev. B., 84, 020101 (2011). https://doi.org/10.1103/PhysRevB.84.020101

[46] Liu J.M., Li Q.C., Gao X.S., Yang Y., Zhou X.H., Chen X.Y., Liu Z.G., Phys. Rev. B: Condens. Matter., 66, 054416 (2002). https://doi.org/10.1103/PhysRevB.66.054416

[47] Liu L., Zhang T., Qi Q., Zhang L., Chen W., Xu B., Solid-State Electron., 51, 1029 (2007). https://doi.org/10.1016/j.sse.2007.05.016

[48] Liu X., Ji H., Gu Y., Xu M., Mat. Scien. and Eng. B., 133, 98 (2006).

https://doi.org/10.1016/j.mseb.2006.06.027

[49] March A., Z. Kristallogr., 81, 285 (1932). https://doi.org/10.1524/zkri.1932.81.1.285

[50] Maxwell J.C., Electricity and Magnetism, Oxford University Press, London (1973).

[51] Meadowcraft D.B., Wimmer J.M., Amer.Ceram. Soc.Bull., 58, 610 (1979).

[52] Momma K., Izumi F., VESTA: a three-dimensional visualization system for electronic and structural analysis. J. Appl. Crystallogr., 41, 653 (2008). https://doi.org/10.1107/S0021889808012016

[53] Moser A., Rettner C.T., Best M.E., Fullerton E.E., Weller D., Parker M., Doerner M.F., IEEE Trans Magn., 36, 2137 (2000). https://doi.org/10.1109/20.908333

[54] Moulson A.J., Herbert J.M., Electroceramics: Materials, properties and applications, 2nd edition, Wiley, New York (2003). https://doi.org/10.1002/0470867965

[55] Mukhopadhyay K., Mahapatra A.S., Chakrabarti P.K., J. Magn. Magn. Mater., 329, 133 (2013). https://doi.org/10.1016/j.jmmm.2012.09.063

[56] Nan C.W., Bichurin M.I., Dong S., Viehland D., Srinivasan G., J. Appl. Phys., 103, 031101 (2008). https://doi.org/10.1063/1.2836410

[57] Nguyen A.T., Knurova M.V., Nguyen T.M., Mittova V.O., Mittova I.Ya., Nanosyst.: Physics, Chem. Math., 5, 692 (2014).

[58] Parkin S.S.P., Roche K.P., Samant M.G., Rice P.M., Beyers R.B., Scheuerlein R. E., O'sullivan E.J., Brown S.L, Bucchigano J., Abraham D.W., Lu Yu, Rooks M., Trouilloud P.L., Wanner R.A., Gallagher W.J., J. Appl. Phys., 85, 5828 (1999). https://doi.org/10.1063/1.369932

[59] Patankar K.K., Dombale P.D., Mathe V.L., Patil S.A., Patil R.N., Mater. Sci. Eng. B., 8, 53 (2001). https://doi.org/10.1016/S0921-5107(01)00695-X

[60] Petricek V., Dusek M., Palatinus L., Kristallogr Z, Crystallographic Computing System JANA2006: General features., 229, 345 (2014). https://doi.org/10.1515/zkri-2014-1737

[61] Phokha S., Hunpratup S., Pinitsoontorn S., Putasaeng B., Rujirawat S., Maensiri S., Mater. Res. Bull., 67, 118 (2015). https://doi.org/10.1016/j.materresbull.2015.03.008

[62] Ponpian N., Balaya P., Narayansamy A., J. Phys. ,Conden. Mater., 14, 3221 (2002). https://doi.org/10.1088/0953-8984/14/12/311

[63] Rick U., J. Am. Ceram. Soc., 90, 3326 (2007). https://doi.org/10.1111/j.1551-

2916.2007.01881.x

[64] Rietveld H.M., J. Appl. Crystallogr., 2, 65 (1969).
https://doi.org/10.1107/S0021889869006558

[65] Sakai M., Masuno A., Kan D., Hashisaka M., Takata K., Azuma M., Takano M.,
Shimakawa Y., Appl. Phys. Lett., 90, 072903 (2007).
https://doi.org/10.1063/1.2539575

[66] Sakata M., Sato M., Acta Cryst., A46, 263 (1990).
https://doi.org/10.1107/S0108767389012377

[67] Saravanan R., Grain software (Personal communication) (2008), http://phymat.in/.

[68] Sawyer C.B., Tower C.H., Phys. Rev., 35, 269 (1930).
https://doi.org/10.1103/PhysRev.35.269

[69] Schmid H, Ferroelectricity, 162 317 (1994).
https://doi.org/10.1080/00150199408245120

[70] Shikha P., Kang T.S., Randhawa B.S., J. Alloys Compd., 625, 336 (2015).
https://doi.org/10.1016/j.jallcom.2014.11.074

[71] Skoog D.A., Holler F.J., Crouch S.R., Principles of Instrumental Analysis. 6th
edition, Thomson Brooks, USA, (2007).

[72] Smolenskii G.A., Fizika segnetoelektrikov, Nauka, Leningrad (1978).

[73] Spaldin N.A., Fiebig M., Mostovoy M., J. Phys.: Condens. Matter. 20,
434203(2008). https://doi.org/10.1088/0953-8984/20/43/434203

[74] Steele B.C., Heinzel A., Nature, 414, 345 (2001). https://doi.org/10.1038/35104620

[75] Stout G.H., Jensen L.H., X-ray structure determination-a practical guide. The
Macmillan Company Collier-Macmillan, London, 217 (1968).

[76] Tang P., Tong Y., Chen H., Cao F., Pan G., Appl. Phys., 13, 340 (2013).
https://doi.org/10.1016/j.cap.2012.08.006

[77] Thirumalairajan S., Girija K., Mastelaro V.R., Ponpandian N., J. Mater. Sci.- Mater.
Electron., 26, 8652 (2015). https://doi.org/10.1007/s10854-015-3540-z

[78] Thompson P., Cox D.E., Hastings J.B., J Appl Crystallogr., 20, 79 (1987).
https://doi.org/10.1107/S0021889887087090

[79] Toa N.N., Saukko S., Lantto V., Phys. B Condens. Matter., 327, 279 (2003).
https://doi.org/10.1016/S0921-4526(02)01764-7

[80] Triyono D., Kafa C.A., Laysandra H., J. Adv. Dielect., 8, 1850036 (2018)

https://doi.org/10.1142/S2010135X18500364

[81] Van Aken B.B., Palstra T.T.M., Filippetti A., Spaldin N.A., Nat. Mater., 3, 164 (2004). https://doi.org/10.1038/nmat1080

[82] Vandeven D., Galy J., Pouchard M., Hagenmuller P., Mat. Res. Bull., 2, 809 (1967). https://doi.org/10.1016/0025-5408(67)90008-6

[83] Vasques C., Kogerler P., Lopez–Quintela M.A., Sanchez R.D., Rivas J., J. Mater. Res., 13, 451 (1998). https://doi.org/10.1557/JMR.1998.0058

[84] Vishwanathan B., Murthy V.R.K., Ferrite Materials, Science and Technology, Narosa, New Delhi (1990).

[85] Waghmare U.V., Spaldin N.A., Kandpal H.C., Seshadri Ram, Phys. Rev. B., 67, 125111 (2003). https://doi.org/10.1103/PhysRevB.67.125111

[86] Wagner K.W., Ann. Phys., 40, 818 (1993).

[87] Wang J., Neoton J.B., Zheng H., Nagarajan V., Ogale S.B., Lie B., Viehland D., Vaithyanathan V., Schlom D.G., Waghmare U.V., Spaldin N.A., Rabe K.M., Wutting M., Ramesh R., Science, 299, 1719 (2003). https://doi.org/10.1126/science.1080615

[88] Wertheim G.K., Butler M.A., West K.W., Buchanan D.N.E., Rev. Sci. Instrum., 45, 1369 (1974). https://doi.org/10.1063/1.1686503

[89] Wesley Burgei, Michael J.Pachan, Herbert Jaeger, Am.J.Phys., 71, 8 (2003). DOI:10.1119/1.1572149 https://doi.org/10.1119/1.1572149

[90] Wolf S.A., Awschalom D.D., Buhrman R.A., Daughton J.M., Molnár S.V., Roukes M.L., Chtchelkanova A.Y., Treger D.M., Science, 294, 1488 (2001). https://doi.org/10.1126/science.1065389

[91] Wood D.L., Tauc J., Phys. Rev. B., 5, 3144 (1972). https://doi.org/10.1103/PhysRevB.5.3144

[92] Xiang X.P., Zhao L.H., Teng B.T., Lang J.J., Hu X., Li T., Fang Y.A., Luo M.F., Lin J.J., Appl. Surf. Sci., 276, 328 (2013). https://doi.org/10.1016/j.apsusc.2013.03.091

[93] Xiao H., Xue C., Song P., Li J., Wang Q., Appl. Surf. Sci., 337, 65 (2015). https://doi.org/10.1016/j.apsusc.2015.02.064

[94] Zhang S.T., Zhang Y., Lu M.H., Du C.L., Chen Y.F., Liu Z.G., Zhu Y.Y., Ming N.B., Appl. Phys. Lett., 88, 162901 (2006). https://doi.org/10.1063/1.2195927

Chapter 2

Results

Abstract

In this work, all the synthesized lanthanum orthoferrite (LFO)-type multiferroics have been characterized using various experimental techniques. The structural characterization has been carried out using powder X-ray diffraction (PXRD). The experimental powder XRD data sets have been subjected to Rietveld refinement through the software program JANA 2006 for the structural properties. The microstructure and surface morphology of the synthesized multiferroics have been examined through the scanning electron microscopy (SEM) technique. The elemental composition of the samples has been analyzed using energy dispersive X-ray spectroscopy (EDS). The energy band gap of the synthesized lanthanum orthoferrite-type multiferroics has been determined from the UV-Vis absorption data obtained using UV-visible (UV-Vis) spectroscopy. Room temperature magnetic measurements have been carried out through vibrating sample magnetometry. The dielectric measurements (Impedance analysis) of the multiferroics have been recorded at room temperature over the wide frequency range. The ferroelectric behavior of the multiferroics has been examined from polarization hysteresis loops measured at room temperature. For the synthesized LFO- type multiferroics, the charge density distribution studies have been done using the high resolution Maximum Entropy Method (MEM) [Collins, 1982], employing software programs PRIMA and. So far, MEM-based charge density studies have been carried out for different materials like non-linear optical materials, lead-free piezoceramic materials, ferrite materials, thermoelectric materials, nano semiconductors and ceramic-ferrite magneto-electric composites, etc.

In this chapter, the experimental results obtained from various characterization techniques and the results obtained from MEM-based charge density distribution studies have been presented.

Keywords

$La_{1-x}Ce_xFeO_3$, $La_{1-x}Zn_xFeO_3$, $La_{1-x}Al_xFeO_3$, $La_{1-x}Sr_xFeO_3$, Dielectric, Ferroelectric

2.1 Introduction

In this work, all the synthesized lanthanum orthoferrite (LFO)-type multiferroics have been characterized using various experimental techniques. The structural characterization has been carried out using powder X-ray diffraction (PXRD). The experimental powder XRD data sets have been subjected to Rietveld refinement [Rietveld, 1969] through the software program JANA 2006 [Petricek et al., 2014] for the structural properties. The microstructure and surface morphology of the synthesized multiferroics have been examined through the scanning electron microscopy (SEM) technique. The elemental composition of the samples has been analyzed using energy dispersive X-ray spectroscopy (EDS). The energy band gap of the synthesized lanthanum orthoferrite-type multiferroics has been determined from the UV-Vis absorption data obtained using UV-visible (UV-Vis) spectroscopy. Room temperature magnetic measurements have been carried out through vibrating sample magnetometry. The dielectric measurements (Impedance analysis) of the multiferroics have been recorded at room temperature over the wide frequency range. The ferroelectric behavior of the multiferroics has been examined from polarization hysteresis loops measured at room temperature. For the synthesized LFO- type multiferroics, the charge density distribution studies have been done using the high resolution Maximum Entropy Method (MEM) [Collins, 1982], employing software programs PRIMA [Izumi, 2002] and VESTA [Momma, 2008]. So far, MEM-based charge density studies have been carried out for different materials like non-linear optical materials [Saravanan et al., 2011], lead-free piezoceramic materials [Sasikumar and Saravanan, 2017], ferrite materials [Kannan et al., 2017], thermoelectric materials [Charles Robert and Saravanan, 2010], nano semiconductors [Saravanakumar et al., 2014] and ceramic-ferrite magneto-electric composites [Meenakshi et al., 2020], etc.

In this chapter, the experimental results obtained from various characterization techniques and the results obtained from MEM-based charge density distribution studies have been presented.

2.2   Structural characterization - Powder X-ray diffraction

The phase purity and crystal structure of the synthesized lanthanum orthoferrite (LFO)-type multiferroics have been confirmed using the powder X-ray diffraction method. The XRD patterns of all the multiferroics have been recorded using Bruker AXS D8 advance X-ray diffractometer at room temperature applying CuKα (λ = 1.54056 Å) radiation in the 2θ range of 5° to 120°, with the step size of 0.02°, slit length 12 mm, primary Soller slit 2.5°, secondary Soller slit 2.5°. In this section, the experimentally observed powder X-ray diffraction patterns of all the synthesized multiferroics and their observed profiles refined through the Rietveld technique [Rietveld, 1969] are presented.In the Figures of fitted XRD profiles, the experimentally observed reflections are shown by 'xx' symbols and the theoretical data are shown by solid lines. The difference between calculated and observed diffraction data is shown at the bottom of each Figure along with Bragg's

Materials Research Forum LLC
https://doi.org/10.21741/9781644902271

positions. The refined structural and profile fitted parameters are tabulated in Tables 2.1, 2.2, 2.3 and 2.4.

2.2.1 Ce-substituted lanthanum orthoferrites - La$_{1-x}$Ce$_x$FeO$_3$

The synthesized La$_{1-x}$Ce$_x$FeO$_3$ multiferroics have been characterized by powder XRD. The observed X-ray diffraction patterns of La$_{1-x}$Ce$_x$FeO$_3$ (x=0.00, 0.03, 0.06, 0.09 & 0.12) multiferroics are shown in Figure 2.1(a). The enlarged XRD peak corresponding to (121) crystallographic plane is shown in Figure 2.1(b). The fitted XRD profiles of La$_{1-x}$Ce$_x$FeO$_3$(x=0.00, 0.03, 0.06, 0.09 & 0.12) multiferroics obtained through Rietveld refinement technique [Rietveld, 1969] using JANA 2006 [Petricek et al., 2014] software program are presented in Figures 2.2(a)-(e). For the synthesized La$_{1-x}$Ce$_x$FeO$_3$ (x=0.00, 0.03, 0.06, 0.09 & 0.12) multiferroics, the cell dimensions have been refined using the unit cell refinement software program [Holland and Redfern, 1997] in order not to bias the cell dimensions due to other structural parameters used in Rietveld software program [Rietveld, 1969]. Table 2.1 gives the structural parameters refined from the unit cell refinement [Holland and Redfern, 1997] and profile fitted parameters obtained from Rietveld refinement [Rietveld, 1969].

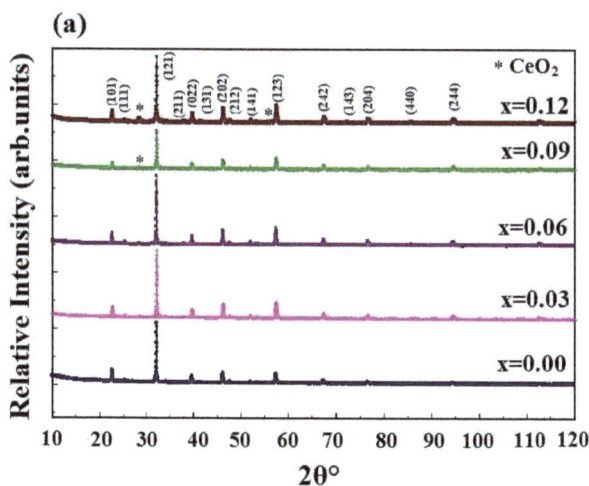

Figure 2.1(a) *Observed XRD patterns of La$_{1-x}$Ce$_x$FeO$_3$ (x=0.00, 0.03, 0.06, 0.09 & 0.12) multiferroics*

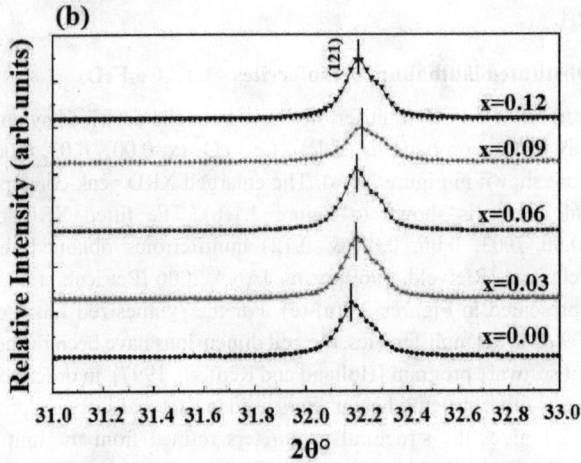

Figure 2.1(b) Enlarged XRD peak of (121) crystallographic plane of $La_{1-x}Ce_xFeO_3$
(x=0.00, 0.03, 0.06, 0.09 & 0.12) multiferroics

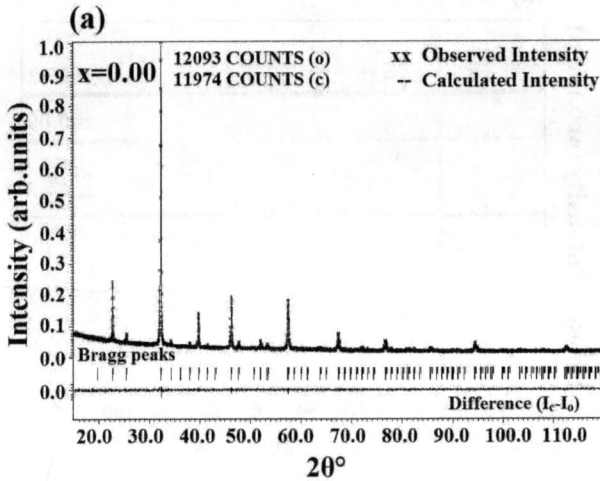

Figure 2.2(a) Fitted powder XRD profile of $La_{1-x}Ce_xFeO_3$, x=0.00

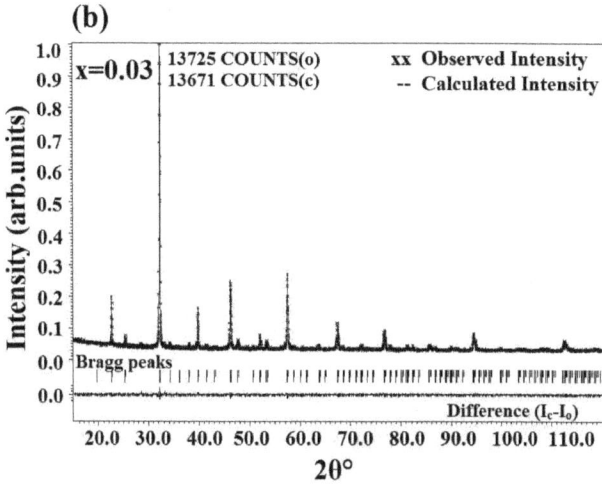

Figure 2.2(b) Fitted powder XRD profile of $La_{1-x}Ce_xFeO_3$, x=0.03

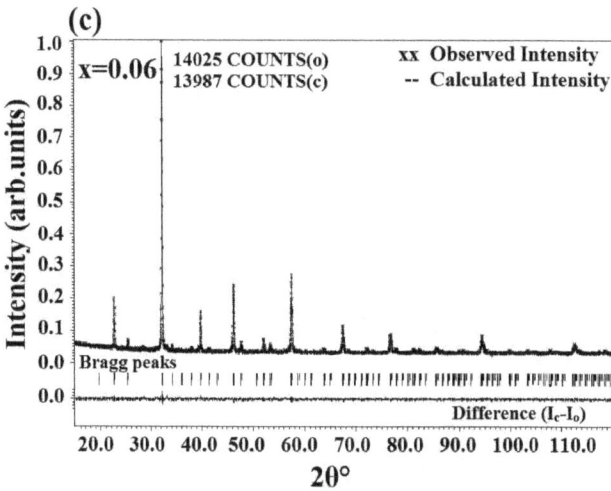

Figure 2.2(c) Fitted powder XRD profile of $La_{1-x}Ce_xFeO_3$, x=0.06

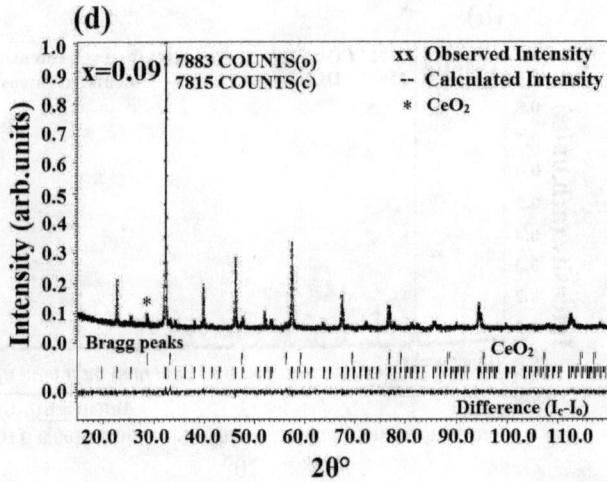

Figure 2.2(d) *Fitted powder XRD profile of $La_{1-x}Ce_xFeO_3$, x=0.09*

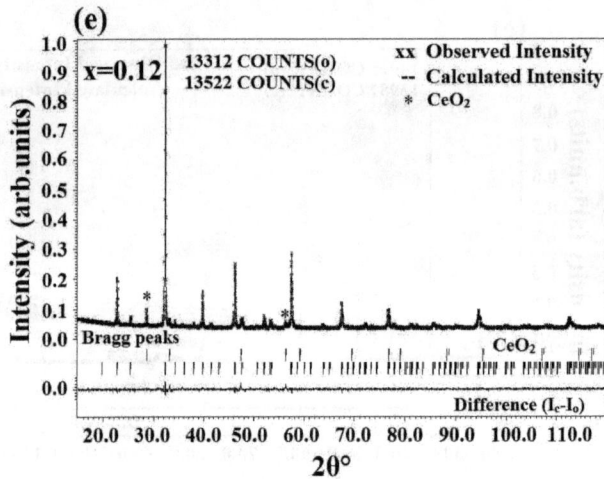

Figure 2.2(e) *Fitted powder XRD profile of $La_{1-x}Ce_xFeO_3$, x=0.12*

Table 2.1 *Refined structural and profile fitted parameters of* $La_{1-x}Ce_xFeO_3$ *multiferroics*

Parameters	x=0.00	x=0.03	x=0.06	x=0.09	x=0.12
*a (Å)	5.5633(27)	5.5619(19)	5.5617(28)	5.5616(35)	5.5612(30)
*b (Å)	7.8546(23)	7.8543(5)	7.8521(23)	7.8485(29)	7.8551(25)
*c (Å)	5.5564(9)	5.5513(6)	5.5512(10)	5.5528(12)	5.5510(10)
*α=β=γ(∘)	90	90	90	90	90
*V (Å³)	242.80(7)	242.51(5)	242.43(9)	242.38(1)	242.49(7)
Robs (%)	2.08	2.40	2.20	2.48	2.49
Rp (%)	3.85	3.63	3.54	3.80	3.61
GOF	1.02	1.10	1.09	1.06	1.19
F(000)	428	428	428	428	428

*a,*b,*c,*α,*β,*γ and *V - Cell parameters and cell volume obtained from unit cellrefinement [Holland and Redfern, 1997]

R_{obs} - Reliability index for observed structure factorsR_P - Reliability index for profile

GOF - Goodness of fit

$F_{(000)}$ - Number of electrons in the unit cell obtained through Rietveldrefinement [Rietveld, 1969]

$F_{(000)}$ is same for all samples: Atomic number of Ce is 58 and that of La is 57.

2.2.2   Zn-substituted lanthanum orthoferrites - $La_{1-x}Zn_xFeO_3$

Figure 2.3(a) shows the observed X-ray diffraction patterns of $La_{1-x}Zn_xFeO_3$ (x=0.00, 0.05, 0.15 & 0.25) multiferroics. Figure 2.3(b) shows the enlarged XRD peak corresponding to (121) crystallographic plane. The Rietveld [Rietveld, 1969] fitted profiles of $La_{1-x}Zn_xFeO_3$ (x=0.00, 0.05, 0.15 & 0.25) multiferroics are shown in Figures 2.4(a)-(d). For the synthesized $La_{1-x}Zn_xFeO_3$ (x=0.00, 0.05, 0.15 & 0.25) multiferroics, the cell dimensions have been refined using the unit cell refinement software program [Holland and Redfern, 1997] in order not to bias the cell dimensions due to other structural parameters used in Rietveld software program [Rietveld, 1969]. The structural parameters obtained from the unit cell refinement method [Holland and Redfern, 1997] and the profile fitted parameters extracted from the Rietveld refinement method [Rietveld, 1969] are given in Table 2.2.

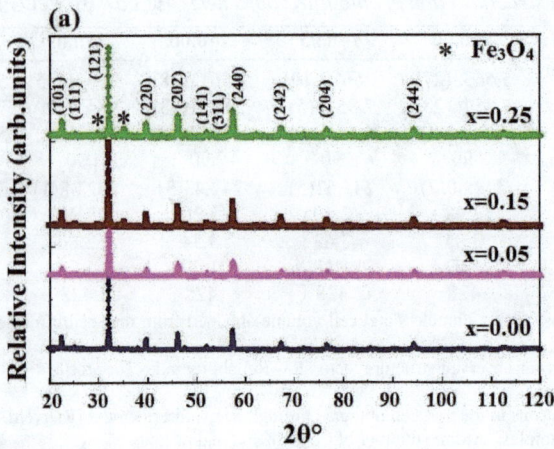

Figure 2.3(a) *Observed XRD patterns of $La_{1-x}Zn_xFeO_3$ (x=0.00, 0.05, 0.15 & 0.25) multiferroics*

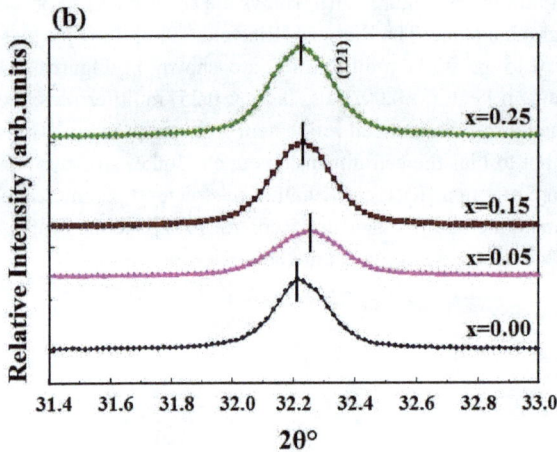

Figure 23(b) *Enlarged XRD peak of (121) crystallographic plane of $La_{1-x}Zn_xFeO_3$*
(x=0.00, 0.05, 0.15 & 0.25) multiferroics

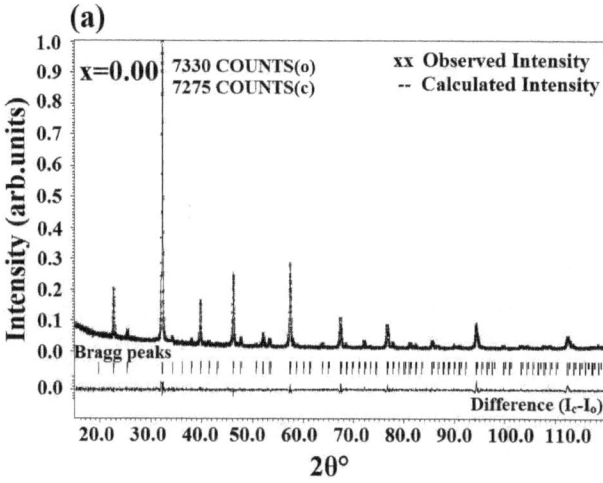

Figure 2.4(a) *Fitted powder XRD profile of* $La_{1-x}Zn_xFeO_3$, *x=0.00*

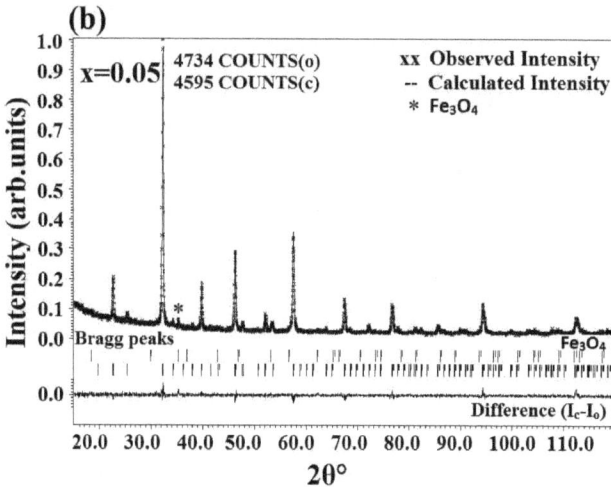

Figure 2.4(b) *Fitted powder XRD profile of* $La_{1-x}Zn_xFeO_3$, *x=0.05*

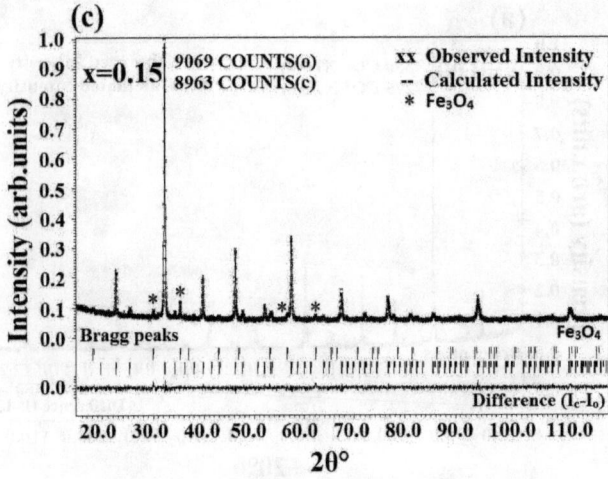

Figure 2.4(c) Fitted powder XRD profile of $La_{1-x}Zn_xFeO_3$, x=0.15

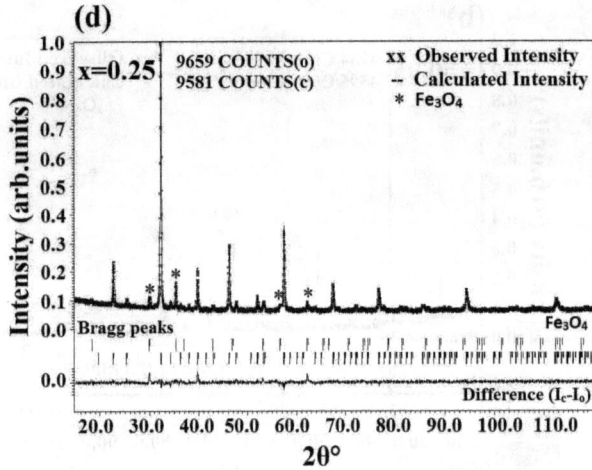

Figure 2.4(d) Fitted powder XRD profile of $La_{1-x}Zn_xFeO_3$, x=0.25

Table 2.2 *Refined structural and profile fitted parameters of* $La_{1-x}Zn_xFeO_3$ *multiferroics*

Parameters	x=0.00	x=0.05	x=0.15	x=0.25
$^*a$ (Å)	5.5629(2)	5.5683(9)	5.5645(4)	5.5655(3)
$^*b$ (Å)	7.8551(1)	7.8486(7)	7.8553(3)	7.8531(2)
$^*c$ (Å)	5.5531(1)	5.5481(6)	5.5507(3)	5.5531(2)
$^*\alpha=\beta=\gamma(°)$	90	90	90	90
$^*V$ (Å^3)	242.66(4)	242.47(3)	242.63(9)	242.71(7)
Robs (%)	2.32	2.49	3.20	3.71
R$_p$(%)	6.16	6.14	3.51	3.60
GOF	1.22	1.14	1.15	1.34
F$_{(000)}$	428	423	412	396

$^*a,^*b,^*c,^*\alpha,^*\beta,^*\gamma$ and $^*V$ - Cell parameters and cell volume obtained from unit cellrefinement [Holland and Redfern, 1997]

R$_{obs}$ - Reliability index for observed structure factorsR$_P$ - Reliability index for profile

GOF - Goodness of fit

F$_{(000)}$ - Number of electrons in the unit cell obtained through Rietveldrefinement [Rietveld, 1969]

2.2.3   Al-substituted lanthanum orthoferrites -La$_{1-x}$Al$_x$FeO$_3$

The observed X-ray diffraction patterns of La$_{1-x}$Al$_x$FeO$_3$ (x=0.05, 0.15 & 0.25) multiferroics are given in Figure 2.5(a). The enlarged XRD peak corresponding to (121) crystallographic plane for all the Al concentrations is shown in Figure 2.5(b). Figures 2.6 (a)-(c) show the fitted Rietveld [Rietveld, 1969] refined XRD profiles of La$_{1-x}$Al$_x$FeO$_3$ (x=0.05, 0.15 & 0.25) multiferroics. The structural and profile fitted parameters extracted from the Rietveld refinement method [Rietveld, 1969] are given in Table 2.3.

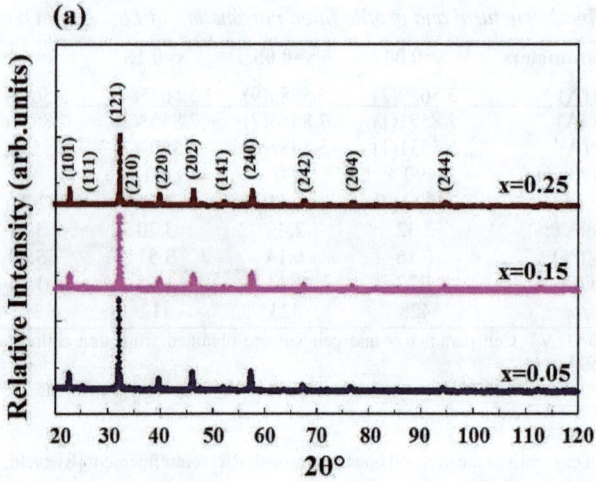

Figure 2.5(a) *Observed XRD patterns of La$_{1-x}$Al$_x$FeO$_3$ (x=0.05, 0.15 & 0.25) multiferroics*

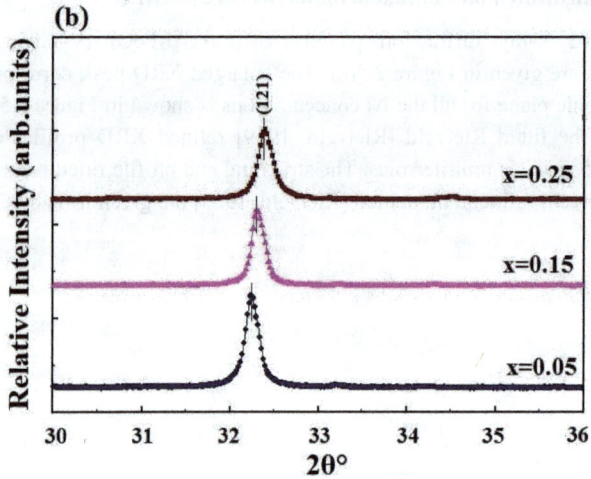

Figure 2.5(b) *Enlarged XRD peak of (121) crystallographic plane of La$_{1-x}$Al$_x$FeO$_3$ (x=0.05, 0.15 & 0.25) multiferroics*

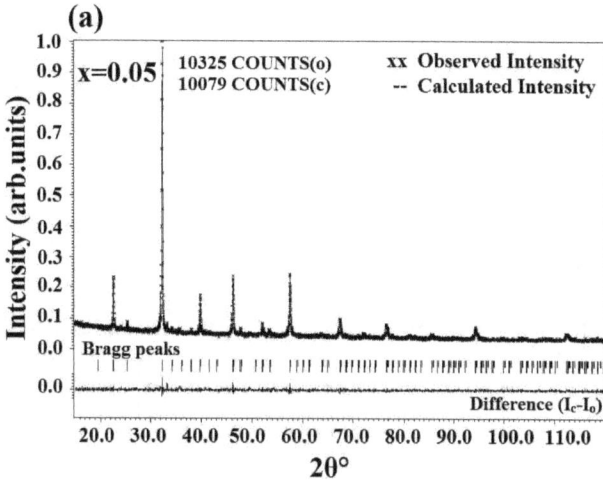

Figure 2.6(a) *Fitted powder XRD profile of La$_{1-x}$Al$_x$FeO$_3$, x=0.05*

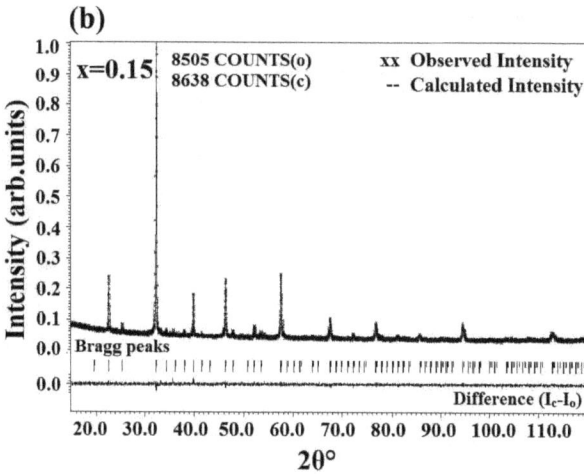

Figure 2.6(b) *Fitted powder XRD profile of La$_{1-x}$Al$_x$FeO$_3$, x=0.15*

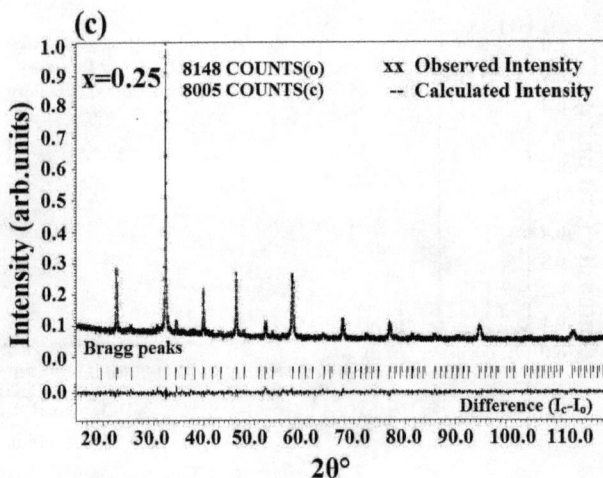

Figure 2.6(c) *Fitted powder XRD profile of $La_{1-x}Al_xFeO_3$, $x=0.25$*

Table 2.3 *Refined structural and profile fitted parameters of $La_{1-x}Al_xFeO_3$ multiferroics*

Parameters	x=0.05	x=0.15	x=0.25
a (Å)	5.5643(4)	5.5567(3)	5.5432(2)
b (Å)	7.8553(11)	7.8493(3)	7.8401(2)
c (Å)	5.5611(6)	5.5577(2)	5.5437(5)
$\alpha=\beta=\gamma(\circ)$	90	90	90
V (Å³)	243.07(2)	242.41(17)	240.93(2)
D (gm/cc)	6.4782(5)	6.1893(4)	5.9189(5)
Robs (%)	2.37	3.81	4.00
R_p (%)	3.68	3.97	3.72
GOF	1.09	1.06	1.16
$F_{(000)}$	419	402	384

V - Cell Volume D - Cell density

R_{obs} - Reliability index for observed structure factors R_P - Reliability index
for profile

GOF - Goodness of fit

$F_{(000)}$ - Number of electrons in the unit cell obtained through Rietveld refinement [Rietveld, 1969]

2.2.4 Sr-substituted lanthanum orthoferrites - $La_{1-x}Sr_xFeO_3$

The observed powder XRD patterns of $La_{1-x}Sr_xFeO_3$ (x=0.05, 0.10, 0.15 & 0.20) multiferroics are given in Figure 2.7(a). The enlarged XRD peak corresponding to (121) crystallographic plane for all the Sr concentrations is shown in Figure 2.7(b). Figures 2.8 (a)-(d) show the fitted powder XRD profiles of $La_{1-x}Sr_xFeO_3$ (x=0.05, 0.10, 0.15 & 0.20) multiferroics. The structural and profile fitted parameters obtained from the Rietveld refinement method [Rietveld, 1969] are tabulated in Table 2.4.

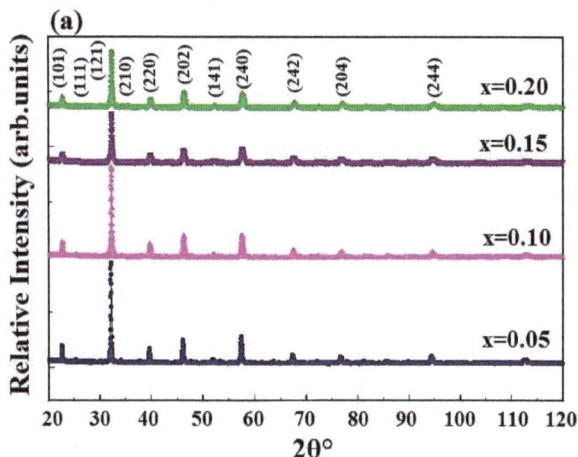

Figure 2.7(a) *Observed XRD patterns of $La_{1-x}Sr_xFeO_3$ (x=0.05, 0.10, 0.15 & 0.20) multiferroics*

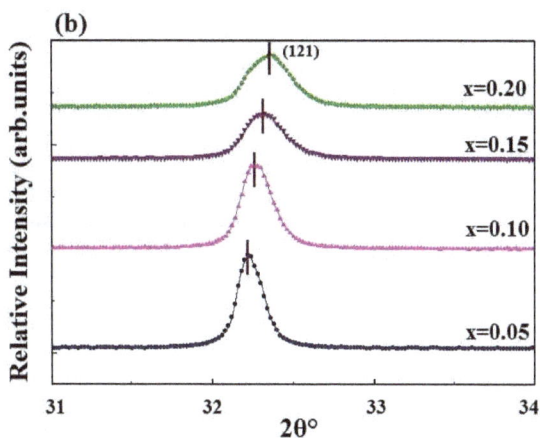

Figure 2.7(b) *Enlarged XRD peak of (121) crystallographic plane of $La_{1-x}Sr_xFeO_3$ (x=0.05, 0.10, 0.15 & 0.20) multiferroics*

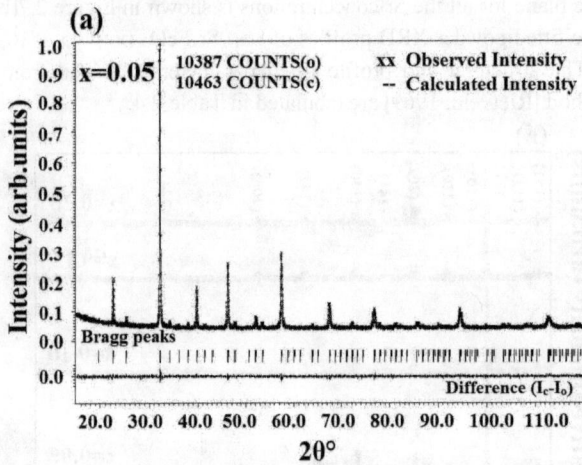

Figure 2.8(a) Fitted powder XRD profile of $La_{1-x}Sr_xFeO_3$, x=0.05

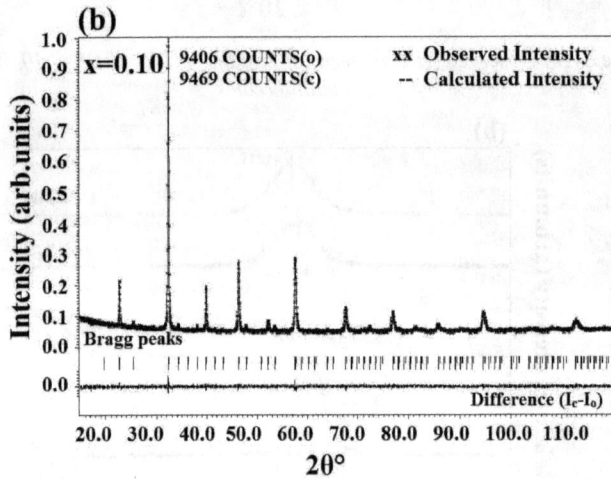

Figure 2.8(b) Fitted powder XRD profile of $La_{1-x}Sr_xFeO_3$, x=0.10

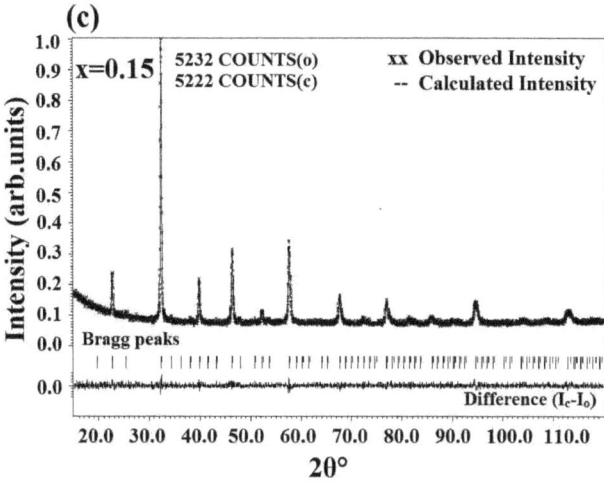

Figure 2.8(c) *Fitted powder XRD profile of La$_{1-x}$Sr$_x$FeO$_3$, x=0.15*

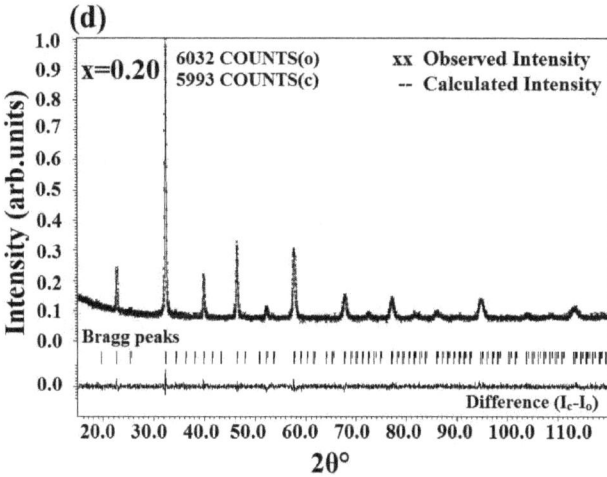

Figure 2.8(d) *Fitted powder XRD profile of La$_{1-x}$Sr$_x$FeO$_3$, x=0.20*

Table 2.4 *Refined structural and profile fitted parameters of La$_{1-x}$Sr$_x$FeO$_3$ multiferroics*

Parameters	x=0.05	x=0.10	x=0.15	x=0.20
a (Å)	5.5551(4)	5.5516(4)	5.5421(3)	5.5313(4)
b (Å)	7.8452(4)	7.8443(6)	7.8410(5)	7.8317(7)
c (Å)	5.5532(3)	5.5507(2)	5.5435(5)	5.5389(3)
α= β= γ (°)	90	90	90	90
V (Å^3)	242.02(3)	241.73(6)	240.90(2)	239.95(7)
D(gm/cc)	6.5896(3)	6.5272(4)	6.4789(5)	6.4337(7)
Robs (%)	2.61	2.27	3.62	2.57
R$_p$ (/%)	3.45	3.45	3.90	3.70
GOF	1.06	1.07	1.08	1.10
F$_{(000)}$	424	420	417	413

V - Cell Volume

D - Cell density

R$_{obs}$ - Reliability index for observed structure factorsR$_P$ - Reliability index for profile

GOF - Goodness of fit

F$_{(000)}$ - Number of electrons in the unit cell obtained through Rietveldrefinement [Rietveld, 1969]

2.3   Microstructure and elemental characterization - SEM/EDS

The surface morphology and microstructure of all the synthesized multiferroics have been analyzed from SEM images. The elemental compositions of all the synthesized multiferroics have been analyzed qualitatively and quantitatively from energy dispersive spectroscopy (EDS) measurements. In this section, the SEM micrographs and EDS spectra of all the synthesized multiferroics are presented. The numerical values of atomic and weight percentages of the various elements present in the samples for all the synthesized multiferroics are tabulated in Tables 2.5, 2.6, 2.7 & 2.8.

2.3.1   Ce-substituted lanthanum orthoferrites - La$_{1-x}$Ce$_x$FeO$_3$

The SEM micrographs of La$_{1-x}$Ce$_x$FeO$_3$ (x=0.00, 0.03, 0.06, 0.09 & 0.12) multiferroics have been recorded with different magnifications (x500, x700, x1500, x7000, x10000). Figures 2.9(a)-(e) illustrate the SEM micrographs of La$_{1-x}$Ce$_x$FeO$_3$ corresponding to the magnification of x7000.

Figures 2.10(a)-(e) show the EDS spectra of La$_{1-x}$Ce$_x$FeO$_3$ (x=0.00, 0.03, 0.06,

0.09 & 0.12) multiferroics. The numerical values of atomic and weight percentages of the constituent elements of La$_{1-x}$Ce$_x$FeO$_3$ multiferroics are given in Table 2.5.

Figure 2.9 *SEM images of La$_{1-x}$Ce$_x$FeO$_3$ multiferroics,* **(a)** *x=0.00,* **(b)** *x=0.03,* **(c)** *x=0.06,* **(d)** *x=0.09 &* **(e)** *x=0.12*

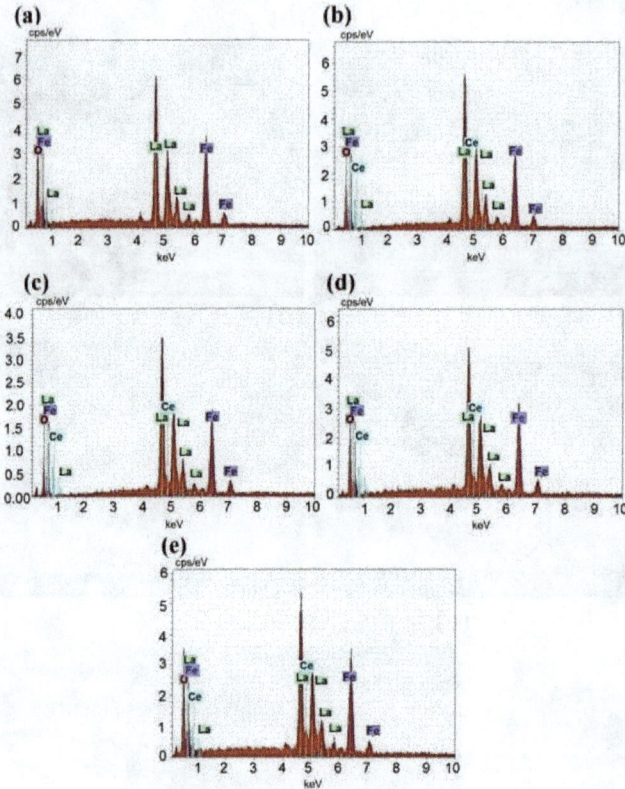

Figure 2.10 *EDS spectra of $La_{1-x}Ce_xFeO_3$ multiferroics, (**a**) x=0.00, (**b**) x=0.03, (**c**) x=0.06, (**d**) x=0.09 & (**e**) x=0.12*

Table 2.5 *EDS elemental compositions of $La_{1-x}Ce_xFeO_3$ multiferroics*

Samples	Atomic (%)				Weight (%)			
	La	Ce	Fe	O	La	Ce	Fe	O
x=0.00	12.73	-	15.51	71.77	46.74	-	22.9	30.36
x=0.03	21.61	0.81	22.5	55.08	57.14	2.17	23.92	16.77
x=0.06	26.44	1.98	27.74	43.84	59.24	4.46	24.98	11.31
x=0.09	15.91	1.46	19.79	62.84	48.83	4.52	24.43	20.22
x=0.12	12.75	1.64	15.76	69.86	44.28	5.75	22.01	27.96

2.3.2   Zn-substituted lanthanum orthoferrites - La$_{1-x}$Zn$_x$FeO$_3$

The SEM micrographs of La$_{1-x}$Zn$_x$FeO$_3$ (x=0.00, 0.05, 0.15 & 0.25) multiferroics have been recorded with different magnifications (x500, x2000, x5000, x7000, x10000, x 15000). Figures 2.11(a)-(d) depict the SEM images of La$_{1-x}$Zn$_x$FeO$_3$ corresponding to x7000 magnification.

Figures 2.12(a)-(d) show the EDS spectra of La$_{1-x}$Zn$_x$FeO$_3$ (x=0.00, 0.05, 0.15 & 0.25) multiferroics. The numerical values of atomic and weight percentages of the various elements present in the synthesized La$_{1-x}$Zn$_x$FeO$_3$ multiferroics are tabulated in Table 2.6.

Figure 2.11 *SEM images of La$_{1-x}$Zn$_x$FeO$_3$ multiferroics, **(a)** x=0.00, **(b)** x=0.05,* *(c) x=0.15 & (d) x=0.25*

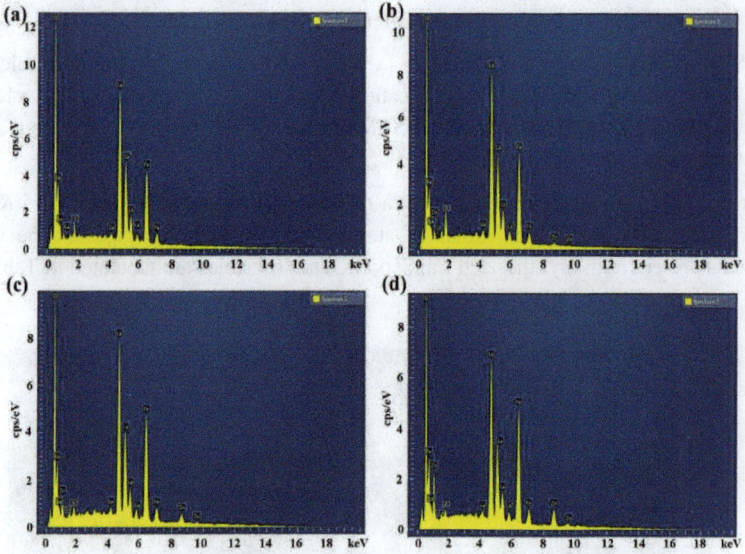

Figure 2.12 EDS spectra of $La_{1-x}Zn_xFeO_3$ multiferroics, **(a)** x=0.00, **(b)** x=0.05, **(c)** x=0.15 & **(d)** x=0.25

Table 2.6 EDS elemental compositions of $La_{1-x}Zn_xFeO_3$ multiferroics

Samples	Atomic (%)				Weight (%)			
	La	Zn	Fe	O	La	Zn	Fe	O
x=0.00	18.51	-	17.23	64.26	56.36	-	21.10	22.54
x=0.05	19.25	0.94	19.58	60.24	55.79	1.28	22.82	20.11
x=0.15	18.84	3.48	20.54	57.13	53.34	4.64	23.38	18.63
x=0.25	16.14	5.45	22.11	56.30	47.36	7.53	26.08	19.03

2.3.3   Al-substituted lanthanum orthoferrites - $La_{1-x}Al_xFeO_3$

The SEM images of $La_{1-x}Al_xFeO_3$ (x=0.05, 0.15 & 0.25) multiferroics have been recorded with various magnifications (x2000, x5000, x10000, x15000). The SEM images of $La_{1-x}Al_xFeO_3$ corresponding to x10000 magnification are given in Figures 2.13(a)-(c).

The EDS spectra of the synthesized $La_{1-x}Al_xFeO_3$ (x=0.05, 0.15 & 0.25) multiferroics are presented in Figures 2.14(a)-(c). Table 2.7 gives the numerical values of atomic and weight percentages of the various elements present in $La_{1-x}Al_xFeO_3$ multiferroics.

Figure 2.13 *SEM images of $La_{1-x}Al_xFeO_3$ multiferroics, **(a)** x=0.05, **(b)** x=0.15 & **(c)** x=0.25*

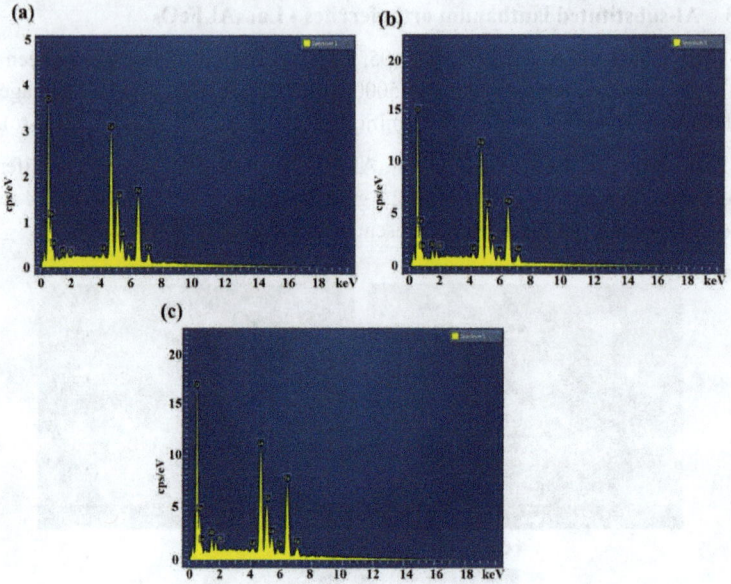

Figure 2.14 *EDS spectra of La$_{1-x}$Al$_x$FeO$_3$ multiferroics,* **(a)** *x=0.05,* **(b)** *x=0.15 &* **(c)** *x=0.25*

Table 2.7 *EDS elemental compositions of La$_{1-x}$Al$_x$FeO$_3$ multiferroics*

Samples	Atomic (%)				Weight (%)			
	La	**Al**	**Fe**	**O**	**La**	**Al**	**Fe**	**O**
x=0.05	20.1	0.52	20.14	59.23	57.23	0.29	23.06	19.42
x=0.15	19.17	2.55	18.43	59.86	56.43	1.46	21.81	20.3
x=0.25	15.89	3.37	20.95	59.78	49.88	2.06	26.45	21.61

2.3.4 Sr-substituted lanthanum orthoferrites - La$_{1-x}$Sr$_x$FeO$_3$

The SEM images of La$_{1-x}$Sr$_x$FeO$_3$ (x=0.05, 0.10, 0.15 & 0.20) multiferroics have been recorded with different magnifications (x500, x1500, x3000, x7000, x10000). The SEM images of La$_{1-x}$Sr$_x$FeO$_3$ corresponding to the magnification x10000 are shown in Figures 2.15(a)-(d).

Figures 2.16(a)-(d) show the EDS spectra of the synthesized La$_{1-x}$Sr$_x$FeO$_3$ (x=0.05, 0.10, 0.15 & 0.20) multiferroics. Table 2.8 gives the numerical values of atomic and weight percentages of the constituent elements of La$_{1-x}$Sr$_x$FeO$_3$ multiferroics.

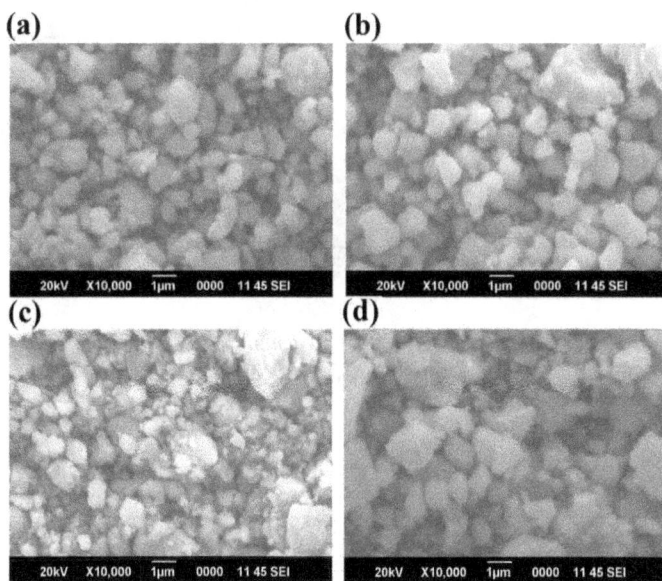

Figure 2.15 *SEM images of La$_{1-x}$Sr$_x$FeO$_3$ multiferroics, **(a)** x=0.05, **(b)** x=0.10, **(c)** x=0.15 & **(d)** x=0.20*

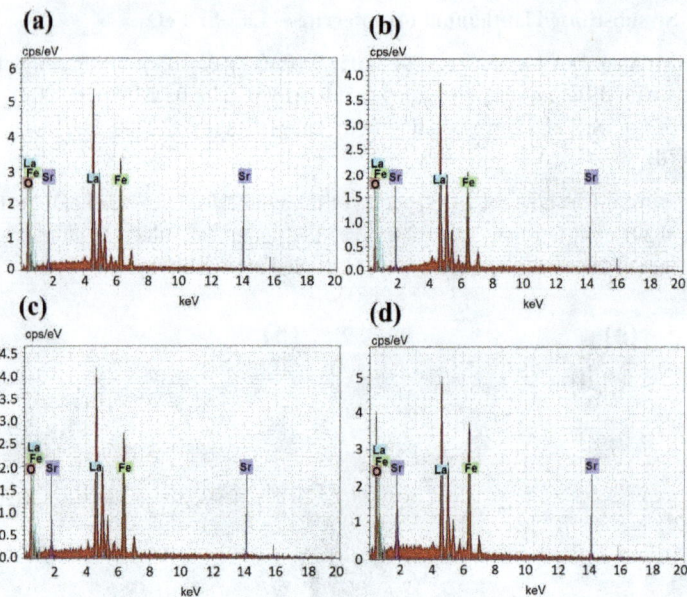

Figure 2.16 *EDS spectra of* $La_{1-x}Sr_xFeO_3$ *multiferroics,* **(a)** *x=0.05,* **(b)** *x=0.10,* **(c)** *x=0.15* & **(d)** *x=0.20*

Table 2.8 *EDS elemental compositions of* $La_{1-x}Sr_xFeO_3$ *multiferroics*

Samples	Atomic (%)				Weight (%)			
	La	**Sr**	**Fe**	**O**	**La**	**Sr**	**Fe**	**O**
x=0.05	16.11	1.38	19.69	62.81	50.14	2.71	24.64	22.51
x=0.10	24.81	2.89	25.29	47.01	58.77	4.32	24.09	12.83
x=0.15	12.42	3.76	16.24	67.58	42.66	8.16	22.44	26.75
x=0.20	8.23	3.90	12.36	75.51	33.80	10.09	20.41	35.70

2.4 Charge density studies-Maximum Entropy Method

The charge density distribution study is an essential part of materials characterization. The accurate electronic structure, charge density distribution and inter- atomic chemical bonding in the unit cell of lanthanum orthoferrite (LFO)-type multiferroics have been analyzed by Maximum Entropy Method (MEM) [Collins, 1982] which is packaged by the software program PRIMA [Izumi and Dilanien, 2002]. In this work, the charge density results are visualized by the visualization software program VESTA [Momma and Izumi, 2008]. The bond angles and bond lengths in the Fe-O sub-lattice of all the synthesized multiferroics are tabulated in Tables 2.9, 2.11, 2.13 & 2.15. The bond lengths and bond critical point charge densities of Fe-O1 and La-O1 bonds in the synthesized LFO-type multiferroics are given in Tables 2.10, 2.12, 2.14 & 2.16.

2.4.1 Ce-substituted lanthanum orthoferrites - $La_{1-x}Ce_xFeO_3$

The three dimensional charge density distribution in the unit cell of $La_{1-x}Ce_xFeO_3$ (x=0.00, 0.03, 0.06, 0.09 & 0.12) multiferroics is shown in Figures 2.17(a)-(e). Figure

2.18 illustrates the three dimensional unit cell of $La_{1-x}Ce_xFeO_3$ (x=0.00) with FeO_6 octahedra. The three dimensional unit cell of $La_{1-x}Ce_xFeO_3$ (x=0.00) with (100) plane shaded is shown in Figure 2.19(a) and the two dimensional charge density distribution of $La_{1-x}Ce_xFeO_3$ (x=0.00) showing Fe-O1 bond on (100) plane is shown in Figure 2.19(b). The enlarged view of Fe-O1 bond for various Ce concentrations x=0.00, 0.03, 0.06, 0.09 & 0.12 is shown in Figures 2.19 (c), (d), (e), (f) & (g) respectively. The three-dimensional unit cell of $La_{1-x}Ce_xFeO_3$ (x=0.00) with (200) plane shaded is shown in Figure 2.20(a) and the two-dimensional charge density distribution of $La_{1-x}Ce_xFeO_3$ (x=0.00) showing La-O1 bond on (200) plane is shown in Figure 2.20(b). The enlarged view of La-O1 bond for various Ce concentrations x=0.00, 0.03, 0.06, 0.09 & 0.12 is shown in Figures 2.20 (c), (d), (e), (f) & (g) respectively. The one-dimensional charge density profiles along the bonds Fe-O1 and La-O1 are shown in Figures 2.21 and 2.22 respectively. The bond angles and bond lengths in the Fe-O sub-lattice of $La_{1-x}Ce_xFeO_3$ multiferroics are tabulated in Table 2.9. The bond lengths and bond critical point charge densities of Fe-O1 and La-O1 bonds of the synthesized LCFO-type multiferroics are given in Table 2.10.

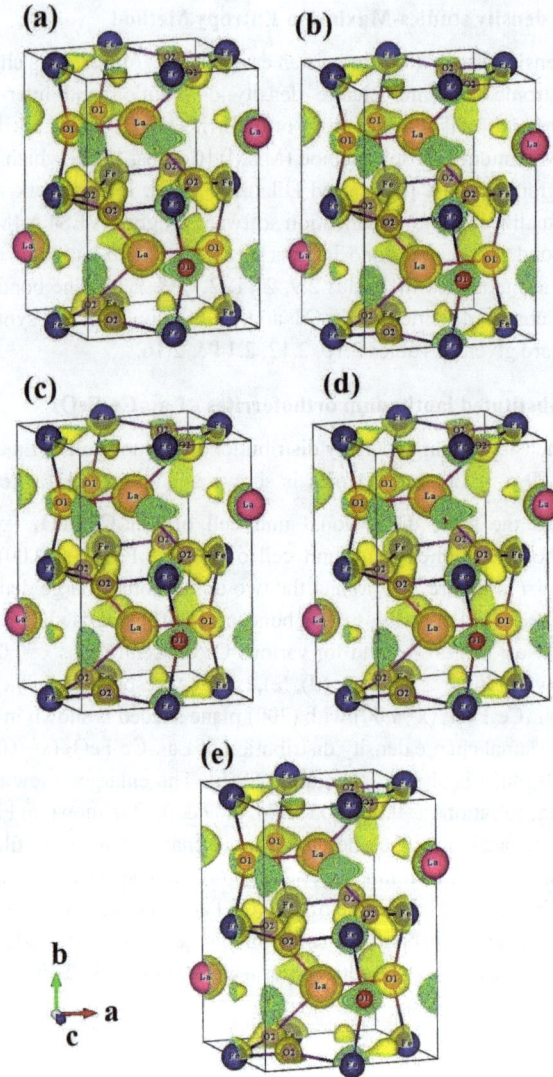

Figure 2.17 *Three-dimensional charge density distribution in the unit cell of*
*$La_{1-x}Ce_xFeO_3$ multiferroics, **(a)** x=0.00, **(b)** x=0.03, **(c)** x=0.06, **(d)** x=0.09 & **(e)** x=0.12*
(Iso-surface level: 3 e/Å³)

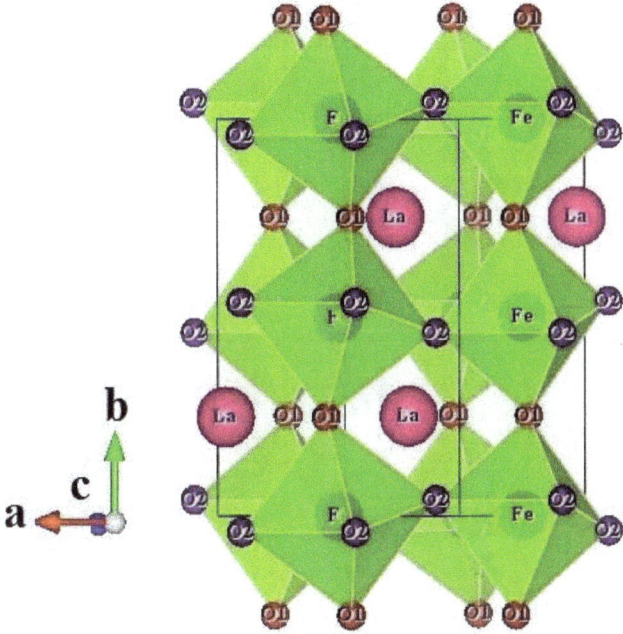

Figure 2.18 *Three-dimensional unit cell of La$_{1-x}$Ce$_x$FeO$_3$ (x=0.00) with FeO$_6$ octahedra*

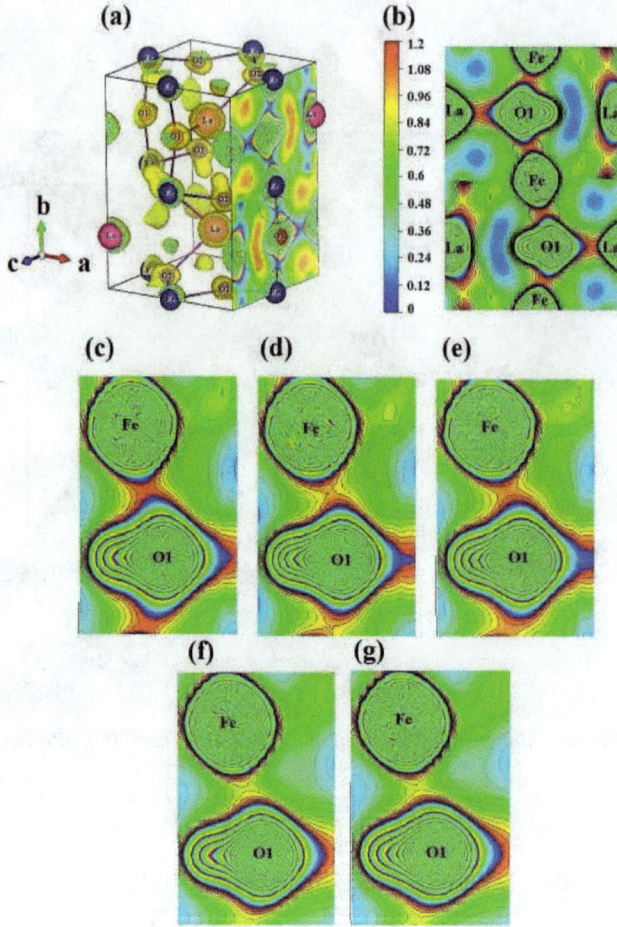

Figure 2.19 (a) *Three-dimensional unit cell of* $La_{1-x}Ce_xFeO_3$ *(x=0.00) with (100) plane shaded.* ***(b)*** *Two-dimensional charge density distribution of* $La_{1-x}Ce_xFeO_3$ *(x=0.00) showing Fe-O1 bond on (100) plane. Enlarged view of Fe-O1 bond,* ***(c)*** *x=0.00,* ***(d)*** *x=0.03,* ***(e)*** *x=0.06,* ***(f)*** *x=0.09 &* ***(g)*** *x=0.12 (Contour range: 0-1.2 e/Å³, contour interval: 0.08 e/Å³)*

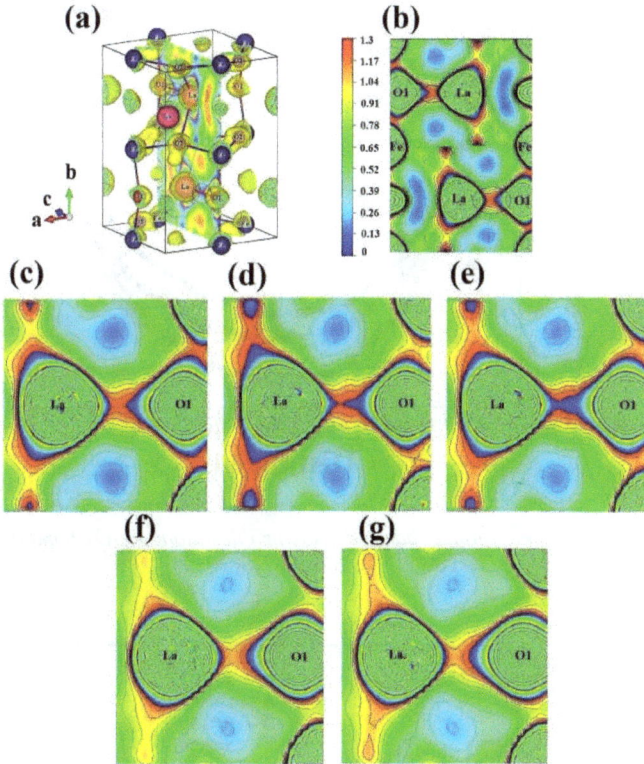

Figure 2.20 (a) *Three-dimensional unit cell of $La_{1-x}Ce_xFeO_3$ (x=0.00) with (200) plane shaded.* ***(b)*** *Two-dimensional charge density distribution of $La_{1-x}Ce_xFeO_3$ (x=0.00) showing La-O1 bond on (200) plane. Enlarged view of La-O1 bond,* ***(c)*** *x=0.00,* ***(d)*** *x=0.03,* ***(e)*** *x=0.06,* ***(f)*** *x=0.09 &* ***(g)*** *x=0.12 (Contour range: 0-1.3 e/Å³, contour interval: 0.09 e/Å³)*

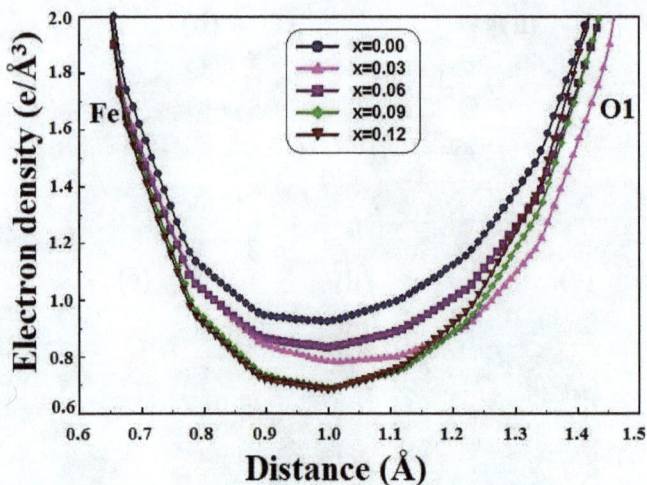

Figure 2.21 *One-dimensional charge density profiles along Fe-O1 bond in* $La_{1-x}Ce_xFeO_3$ *multiferroics*

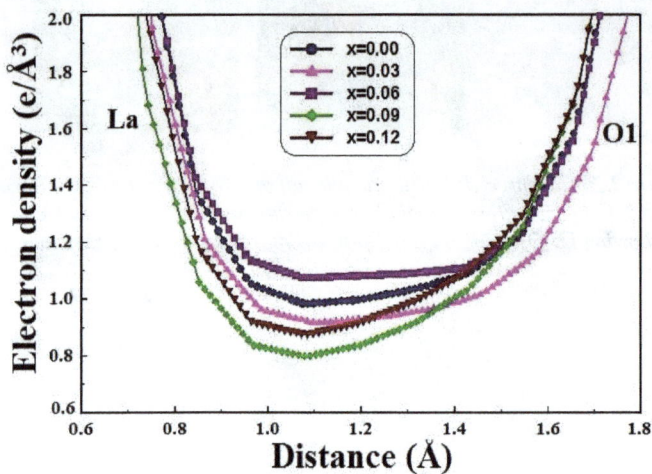

Figure 2.22 *One-dimensional charge density profiles along La-O1 bond in* $La_{1-x}Ce_xFeO_3$ *multiferroics*

Table 2.9 *Bond angles and bond lengths in the Fe-O sub-lattice of $La_{1-x}Al_xFeO_3$ due to $< FeO_6 >$ octahedral twist*

Samples	Bond angle (deg)		Bond length (Å)	
	Fe-O1-Fe	Fe-O2-Fe	Fe-O1	Fe-O2
x=0.00	155.7997	157.2900	2.0081	1.9729
x=0.03	155.5487	157.3569	2.0084	1.9568
x=0.06	155.8081	157.2904	2.0073	1.9720
x=0.09	154.9387	156.9886	2.0074	1.9730
x=0.12	155.5344	156.9582	2.0084	1.9879

Table 2.10 *Bond lengths and bond critical point charge densities of Fe-O1 and La-O1 bonds in $La_{1-x}Ce_xFeO_3$ multiferroics*

Sample	Fe-O1		La-O1	
	Bond length (Å)	Charge density (e/Å³)	Bond length (Å)	Charge density (e/Å³)
x=0.00	2.0081	0.9251	2.4039	0.9811
x=0.03	2.0084	0.7830	2.3985	0.9162
x=0.06	2.0073	0.8340	2.4021	1.0741
x=0.09	2.0074	0.6910	2.3953	0.7983
x=0.12	2.0084	0.6830	2.3998	0.8733

2.4.2 Zn-substituted lanthanum orthoferrites - $La_{1-x}Zn_xFeO_3$

The three-dimensional charge density distribution in the unit cell of $La_{1-x}Zn_xFeO_3$ (x=0.00, 0.05, 0.15 & 0.25) multiferroics is shown in Figures 2.23(a)-(d). Figure 2.24 illustrates the three-dimensional unit cell of $La_{1-x}Zn_xFeO_3$ (x=0.00) with FeO_6 octahedra. The three-dimensional unit cell of $La_{1-x}Zn_xFeO_3$ (x=0.00) with (100) plane shaded is shown in Figure 2.25(a) and the two-dimensional charge density distribution of $La_{1-x}Zn_xFeO_3$ (x=0.00) showing Fe-O1 bond on (100) plane is shown in Figure 2.25(b). The enlarged view of Fe-O1 bond for various Zn concentrations x=0.00, 0.05, 0.15 &

0.25 is shown in Figures 2.25 (c), (d), (e) & (f) respectively. The three-dimensional unit cell of $La_{1-x}Zn_xFeO_3$ (x=0.00) with (200) plane shaded is shown in Figure 2.26(a) and the two-dimensional charge density distribution of $La_{1-x}Zn_xFeO_3$ (x=0.00) showing La-O1 bond on (200) plane is shown in Figure 2.26(b). The enlarged view of La-O1 bond for various Zn concentrations x=0.00, 0.05, 0.15 & 0.25 is shown in Figures 2.26 (c), (d), (e) & (f) respectively. The one-dimensional charge density profiles along the bonds Fe-O1 and La-O1 are shown in Figures 2.27 and 2.28 respectively. The bond angles and bond lengths in the Fe-O sub-lattice of $La_{1-x}Zn_xFeO_3$ multiferroics are tabulated in Table

2.11. The bond lengths and bond critical point charge densities of Fe-O1 and La-O1 bonds of the synthesized $La_{1-x}Zn_xFeO_3$ multiferroics are given in Table 2.12.

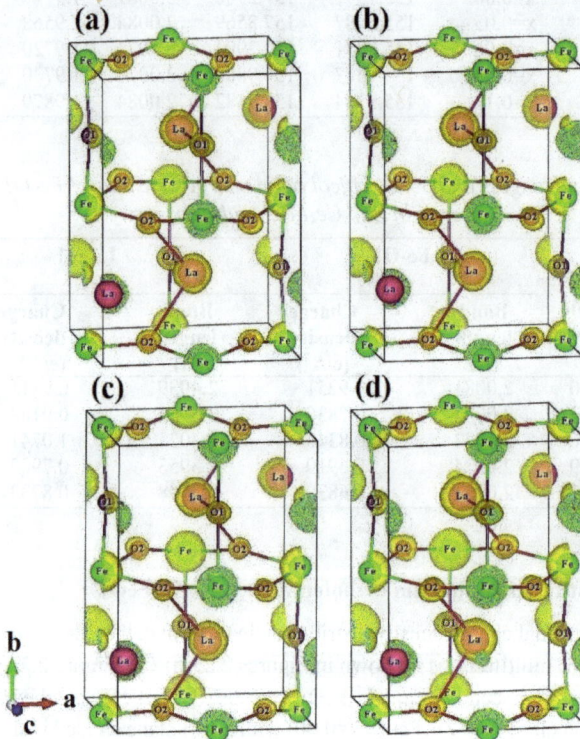

Figure 2.23 *Three-dimensional charge density distribution in the unit cell of* $La_{1-x}Zn_xFeO_3$ *multiferroics,* ***(a)*** *x=0.00,* ***(b)*** *x=0.05,* ***(c)*** *x=0.15 &* ***(d)*** *x=0.25 (Iso-surface level: 3 e/Å³)*

Multiferroic Materials
Materials Research Foundations **140** (2023)

Materials Research Forum LLC
https://doi.org/10.21741/9781644902271

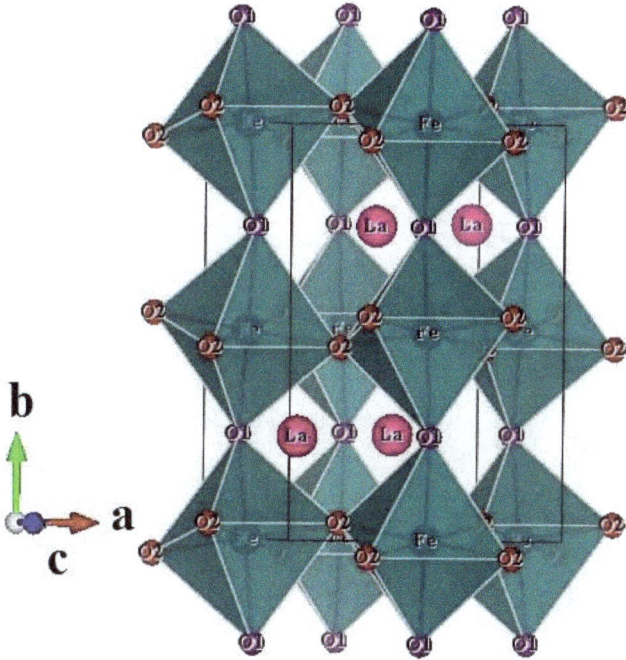

Figure 2.24 *Three-dimensional unit cell of $La_{1-x}Zn_xFeO_3$ (x=0.00) with FeO_6 octahedra*

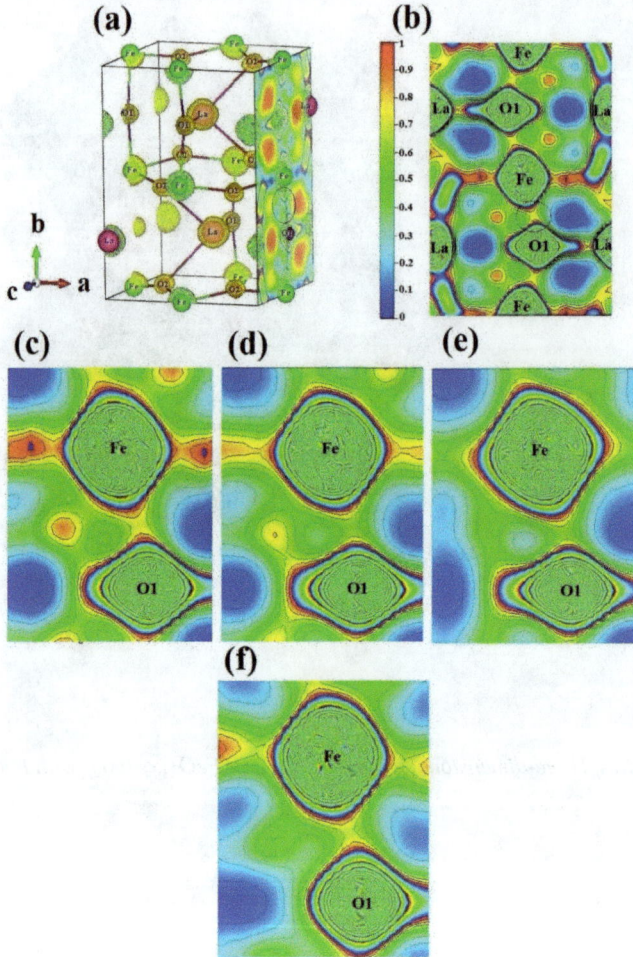

Figure 2.25(a) *Three-dimensional unit cell of La$_{1-x}$Zn$_x$FeO$_3$ (x=0.00) with (100) plane shaded. **(b)** Two-dimensional charge density distribution of La$_{1-x}$Zn$_x$FeO$_3$ (x=0.00) showing Fe-O1 bond on (100) plane. Enlarged view of Fe-O1 bond, **(c)** x=0.00 **(d)** x=0.05, **(e)** x=0.15 & **(f)** x=0.25 (Contour range: 0-1 e/Å^3, contour interval: 0.11 e/Å^3)*

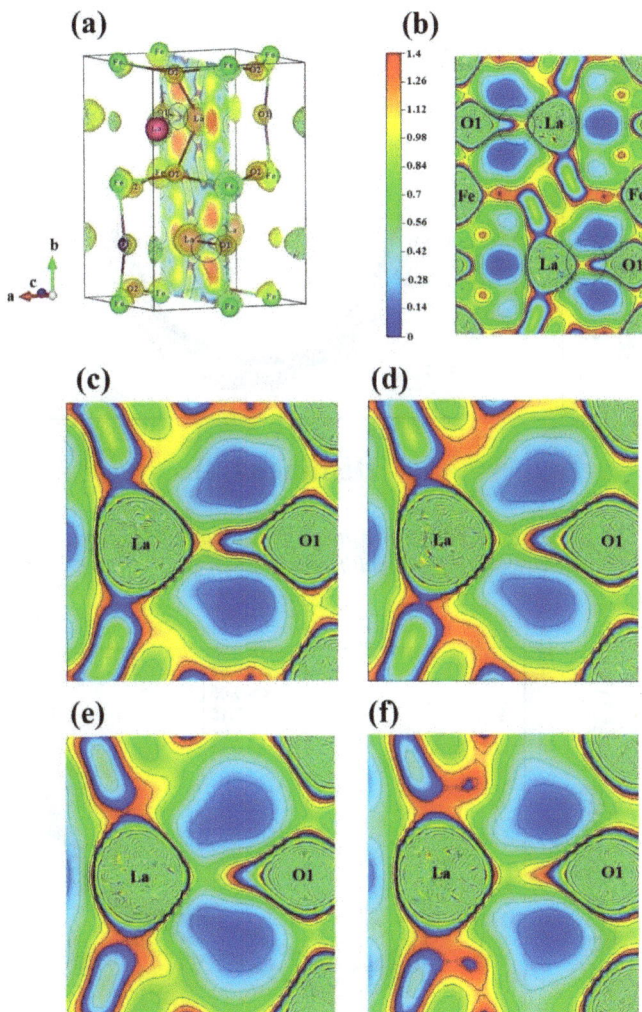

Figure 2.26(a) *Three-dimensional unit cell of $La_{1-x}Zn_xFeO_3$ (x=0.00) with (200) plane shaded.* ***(b)*** *Two-dimensional charge density distribution of $La_{1-x}Zn_xFeO_3$ (x=0.00) showing La-O1 bond on (200) plane. Enlarged view of La-O1 bond,* ***(c)*** *x=0.00,* ***(d)*** *x=0.05,* ***(e)*** *x=0.15 &* ***(f)*** *x=0.25*
(Contour range: 0-1.4 e/Å³, contour interval: 0.09 e/Å³)

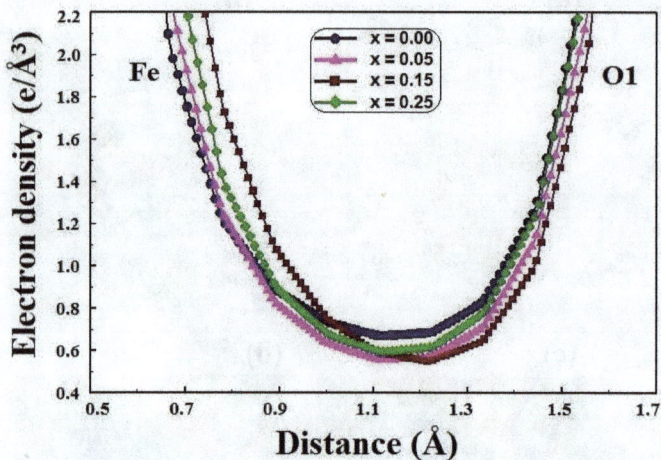

Figure 2.27 *One-dimensional charge density profiles along Fe-O1 bond in $La_{1-x}Zn_xFeO_3$ multiferroics*

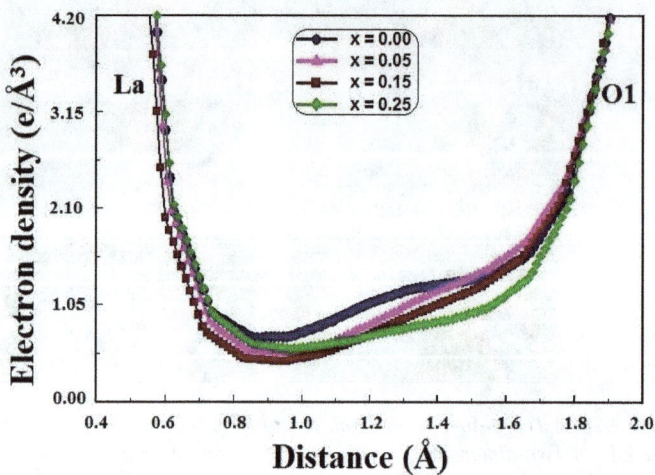

Figure 2.28 *One-dimensional charge density profiles along La-O1 bond in $La_{1-x}Zn_xFeO_3$ multiferroics*

Table 2.11 *Bond angles and bond lengths in the Fe-O sub-lattice of $La_{1-x}Zn_xFeO_3$ due to $<FeO_6>$ octahedral twist*

Samples	Bond angle (deg)	Bond length (Å)	
	Fe–O1–Fe	Fe–O2–Fe	Fe–O1
x=0.00	157.8205	157.1154	2.0020
x=0.05	158.8469	157.4606	1.9991
x=0.15	158.6199	157.1951	1.9990
x=0.25	157.0968	157.0600	2.0077

Table 2.12 *Bond lengths and bond critical point charge densities of Fe-O1 and La-O1 bonds in $La_{1-x}Zn_xFeO_3$ multiferroics*

Sample	Fe-O1		La-O1	
	Bond length (Å)	Charge density (e/Å³)	Bond length (Å)	Charge density (e/Å³)
x=0.00	2.0020	0.6712	2.4453	0.6872
x=0.05	1.9991	0.5553	2.4607	0.5231
x=0.15	1.9990	0.5512	2.4373	0.4326
x=0.25	2.0077	0.5929	2.4368	0.5671

2.4.3 Al-substituted lanthanum orthoferrites -$La_{1-x}Al_xFeO_3$

The three-dimensional charge density distribution in the unit cell of $La_{1-x}Al_xFeO_3$ (x=0.05, 0.15 & 0.25) multiferroics is shown in Figures 2.29 (a)-(c). Figure 2.30 illustrates the three-dimensional unit cell of $La_{1-x}Al_xFeO_3$ (x=0.05) with FeO_6 octahedra. The three-dimensional unit cell of $La_{1-x}Al_xFeO_3$ (x=0.05) with (100) plane shaded is shown in Figure 2.31(a) and the two-dimensional charge density distribution of $La_{1-x}Al_xFeO_3$ (x=0.05) showing Fe-O1 bond on (100) plane is shown in Figure 2.31(b). The enlarged view of Fe-O1 bond for various Al concentrations x=0.05, 0.15 & 0.25 is shown in Figures 2.31 (c), (d) & (e) respectively. The three-dimensional unit cell of $La_{1-x}Al_xFeO_3$ (x=0.05) with (200) plane shaded is shown in Figure 2.32(a) and the two-dimensional charge density distribution of $La_{1-x}Al_xFeO_3$ (x=0.05) showing La-O1 bond on

(200) plane is shown in Figure 2.32(b). The enlarged view of La-O1 bond for various Al concentrations x=0.05, 0.15 & 0.25 is shown in Figures 2.32 (c), (d) & (e) respectively. The one-dimensional charge density profiles along the bonds Fe-O1 and La-O1 are shown

in Figures 2.33 and 2.34 respectively. The bond angles and bond lengths in the Fe-O sub-lattice of $La_{1-x}Al_xFeO_3$ multiferroics are tabulated in Table 2.13. The bond lengths and bond critical point charge densities of Fe-O1 and La-O1 bonds of the synthesized $La_{1-x}Al_xFeO_3$ multiferroics are given in Table 2.14.

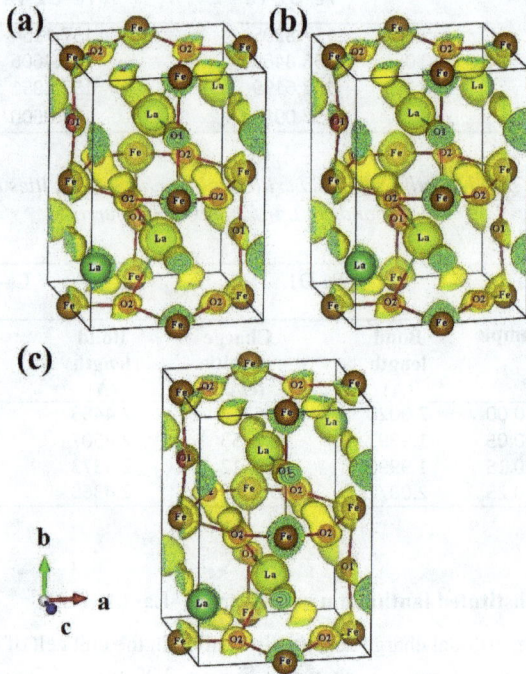

Figure 2.29 *Three-dimensional charge density distribution in the unit cell of $La_{1-x}Al_xFeO_3$ multiferroics, (a) x=0.05, (b) x=0.15 & (c) x=0.25 (Iso-surface level: 3 e/Å³)*

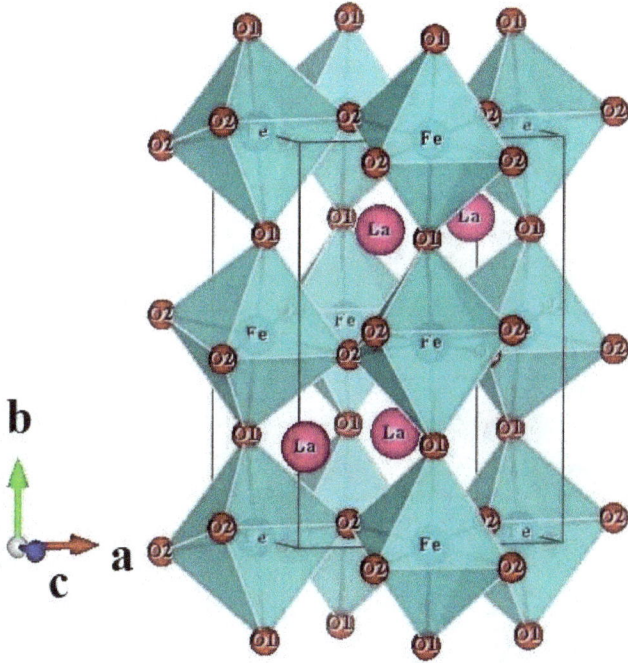

Figure 2.30 *Three-dimensional unit cell of $La_{1-x}Al_xFeO_3$ (x=0.05) with FeO_6 octahedra*

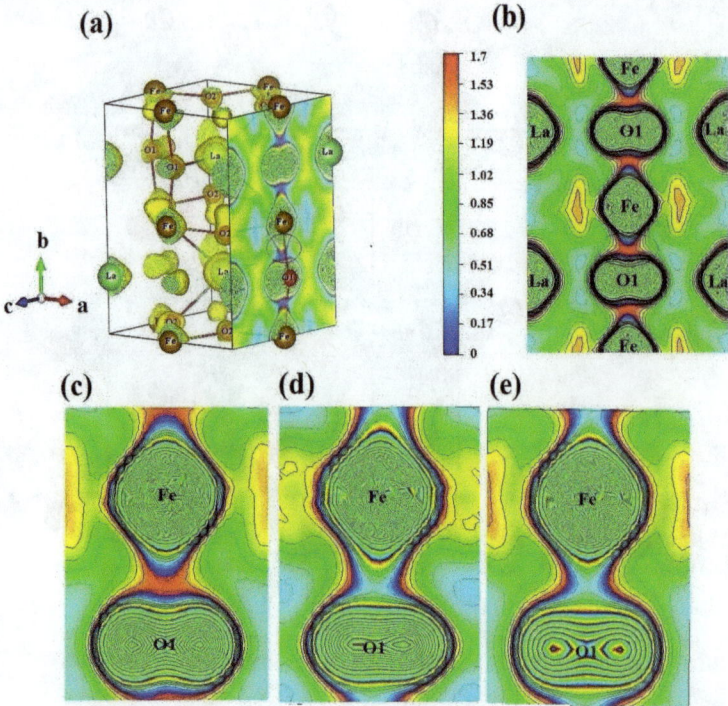

Figure 2.31(a) *Three-dimensional unit cell of* $La_{1-x}Al_xFeO_3$ *(x=0.00) with (100) plane shaded.* **(b)** *Two-dimensional charge density distribution of* $La_{1-x}Al_xFeO_3$ *(x=0.05) showing Fe-O1 bond on (100) plane. Enlarged view of Fe-O1 bond,* **(c)** *x=0.05,* **(d)** *x=0.15 &* **(e)** *x=0.25* *(Contour range: 0-1.7 e/Å³, contour level: 0.12 e/Å³)*

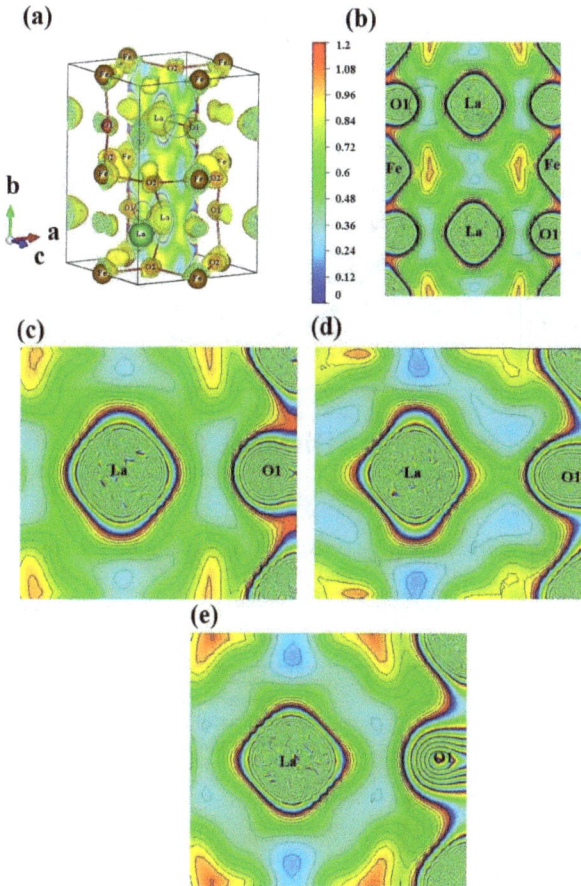

Figure 2.32(a) *Three-dimensional unit cell of La$_{1-x}$Al$_x$FeO$_3$ (x=0.00) with (200) plane shaded.* ***(b)*** *Two-dimensional charge density distribution of La$_{1-x}$Al$_x$FeO$_3$ (x=0.05) showing La-O1 bond on (200) plane. Enlarged view of La-O1 bond,* ***c)*** *x=0.05,* ***(d)*** *x=0.15 &* ***(e)*** *x=0.25 (Contour range: 0-1.2 e/Å^3, contour interval: 0.09 e/Å^3)*

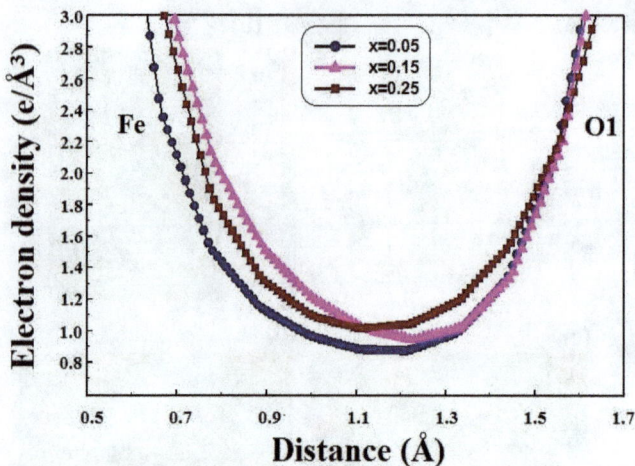

Figure 2.33 One-dimensional charge density profiles along Fe-O1 bond in $La_{1-x}Al_xFeO_3$ multiferroics

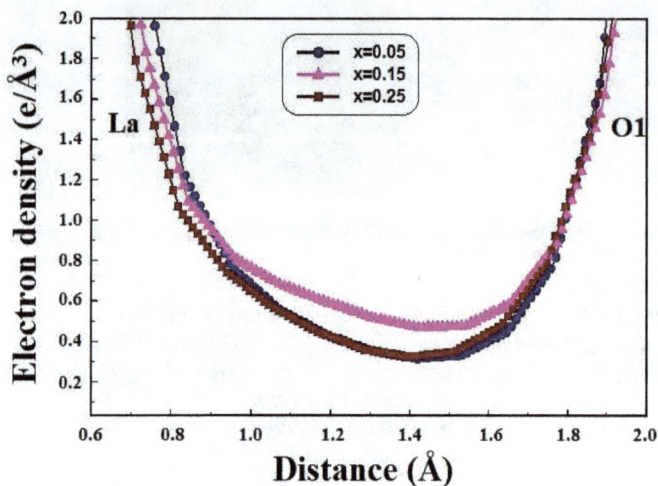

Figure 2.34 One-dimensional charge density profiles along La-O1 bond in $La_{1-x}Al_xFeO_3$ multiferroics

Table 2.13 *Bond angles and bond lengths in the Fe-O sub-lattice of $La_{1-x}Al_xFeO_3$ due to $<$ $FeO_6 >$ octahedral twist*

Samples	Bond angle (deg)		Bond length (Å)	
	Fe–O1–Fe		Fe–O2–Fe	Fe–O1
x=0.05	160.8839	157.8395	1.9914	1.9946
x=0.15	155.7094	156.5067	2.0072	1.9914
x=0.25	157.0381	156.9579	2.0000	2.0006

Table 2.14 *Bond lengths and bond critical point charge densities of Fe-O1 and La-O1 bonds in $La_{1-x}Al_xFeO_3$ multiferroics*

Samples	Fe-O1		La-O1	
	Bond length (Å)	Charge density (e/Å³)	Bond length (Å)	Charge density (e/Å³)
x=0.05	1.9914	0.8760	2.4803	0.3175
x=0.15	2.0072	0.9465	2.3997	0.4699
x=0.25	2.0000	1.0150	2.3968	0.3249

2.4.4 Sr-substituted lanthanum orthoferrites - $La_{1-x}Sr_xFeO_3$

The three-dimensional charge density distribution in the unit cell of $La_{1-x}Sr_xFeO_3$ (x=0.05, 0.10, 0.15 & 0.20) multiferroics is shown in Figures 2.35 (a)-(d). Figure 2.36 illustrates the three-dimensional unit cell of $La_{1-x}Sr_xFeO_3$ (x=0.05) with FeO_6 octahedra. The three-dimensional unit cell of $La_{1-x}Sr_xFeO_3$ (x=0.05) with (100) plane shaded is shown in Figure 2.37(a) and the two-dimensional charge density distribution of $La_{1-x}Sr_xFeO_3$ (x=0.05) showing Fe-O1 bond on (100) plane is shown in Figure 2.37(b). The enlarged view of Fe-O1 bond for various Sr concentrations x=0.05, 0.10, 0.15 & 0.20 is shown in Figures 2.37(c), (d), (e) & (f) respectively. The three-dimensional unit cell of $La_{1-x}Sr_xFeO_3$ (x=0.05) with (200) plane shaded is shown in Figure 2.38(a) and the two-dimensional charge density distribution of $La_{1-x}Sr_xFeO_3$ (x=0.05) showing La-O1 bond on

(200) plane is shown in Figure 2.38(b). The enlarged view of La-O1 bond for various Sr concentrations x=0.05, 0.10, 0.15 & 0.20 is shown in Figures 2.38 (c), (d), (e) & (f) respectively. The one-dimensional charge density profiles along the bonds Fe-O1 and La-O1 are shown in Figures 2.39 and 2.40 respectively. The bond angles and bond lengths in the Fe-O sub-lattice of $La_{1-x}Sr_xFeO_3$ multiferroics are tabulated in Table 2.15. The bond

lengths and bond critical point charge densities of Fe-O1 and La-O1 bonds of the synthesized $La_{1-x}Sr_xFeO_3$ multiferroics are given in Table 2.16.

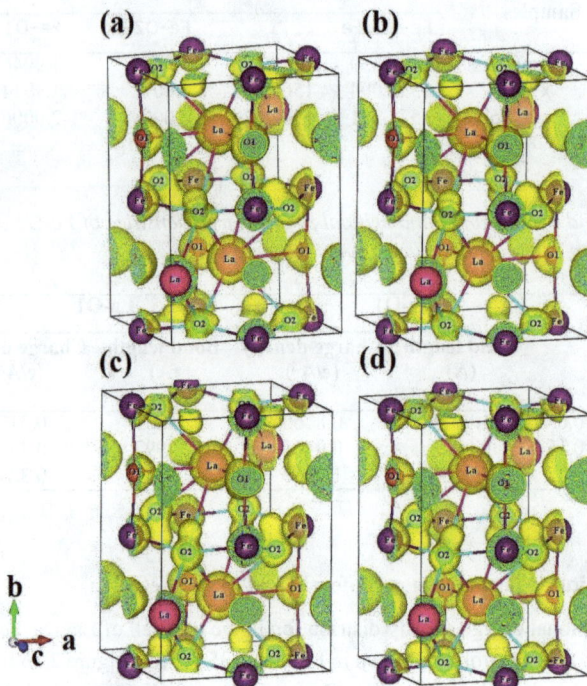

Figure 2.35 *Three-dimensional charge density distribution in the unit cell of $La_{1-x}Sr_xFeO_3$ multiferroics, (a) x=0.05, (b) x=0.10, (c) x=0.15 & (d) x=0.20 (Iso-surface level: 2 e/Å³)*

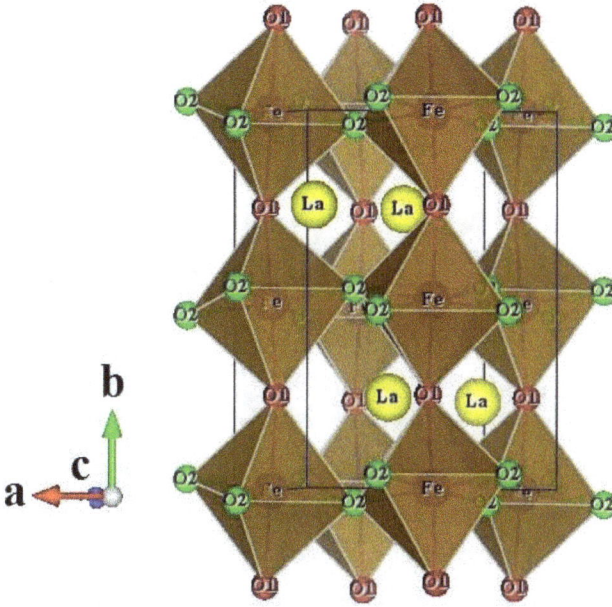

Figure 2.36 Three-dimensional unit cell of $La_{1-x}Sr_xFeO_3$ (x=0.05) with FeO_6 octahedra

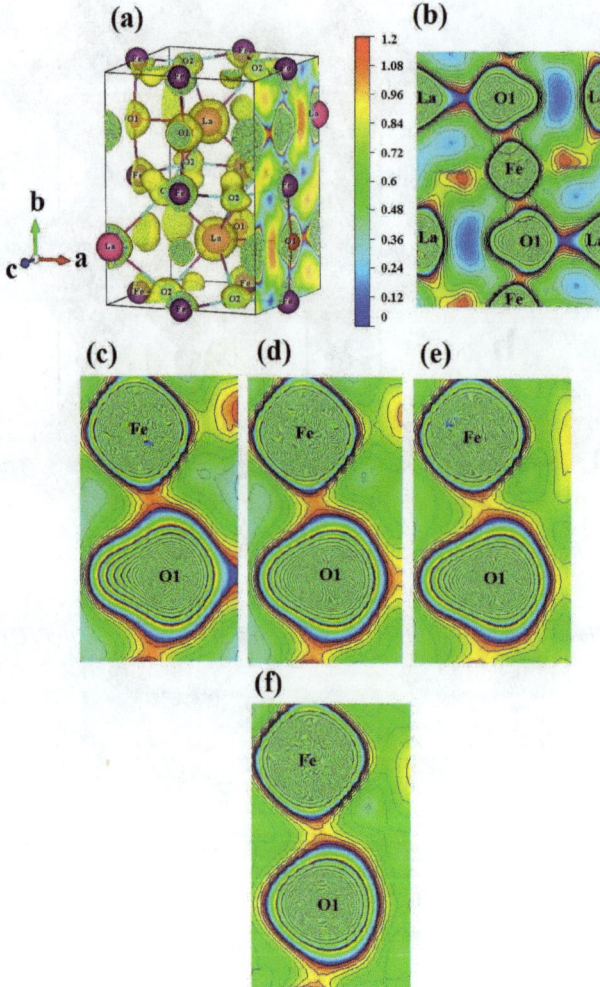

Figure 2.37(a) *Three-dimensional unit cell of $La_{1-x}Sr_xFeO_3$ (x=0.05) with (100) plane shaded. (b) Two-dimensional charge density distribution of $La_{1-x}Sr_xFeO_3$ (x=0.05) showing Fe-O1 bond on (100) plane. Enlarged view of Fe-O1 bond, (c) x=0.05, (d) x=0.10, (e) x=0.15 & (f) x=0.20*
(Contour range: 0-1.2 e/Å³, contour interval: 0.09 e/Å³)

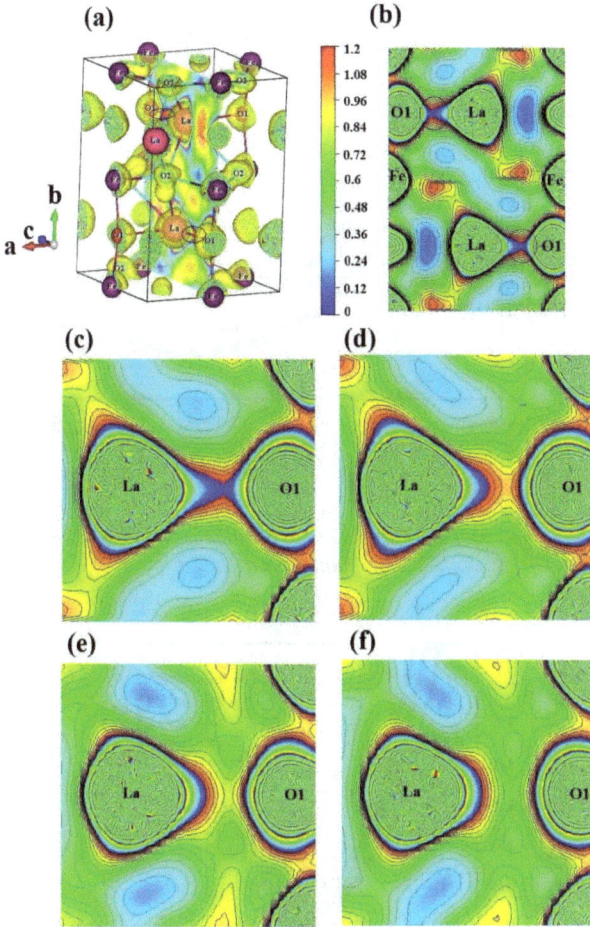

Figure 2.38(a) *Three-dimensional unit cell of La$_{1-x}$Sr$_x$FeO$_3$ (x=0.05) with (200) plane shaded. **(b)** Two-dimensional charge density distribution of La$_{1-x}$Sr$_x$FeO$_3$ (x=0.05) showing La-O1 bond on (200) plane. Enlarged view of La-O1 bond, **(c)** x=0.05, **(d)** x=0.10, **(e)** x=0.15 & **(f)** x=0.20 (Contour range: 0-1.2 e/Å^3, contour interval: 0.09 e/Å^3)*

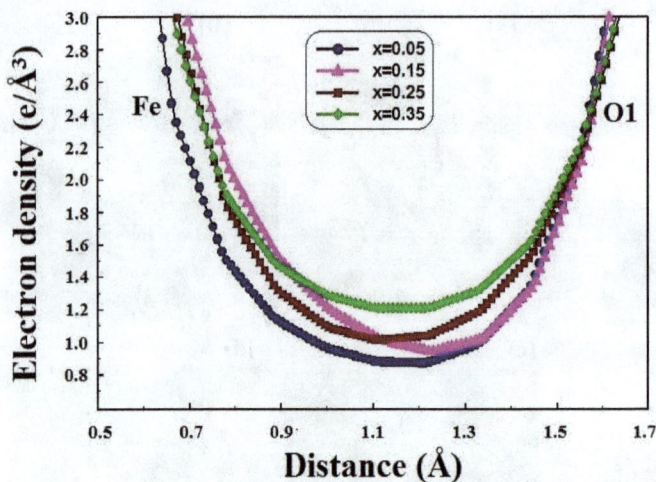

Figure 2.39 *One-dimensional charge density profiles along Fe-O1 bond in La$_{1-x}$Sr$_x$FeO$_3$ multiferroics*

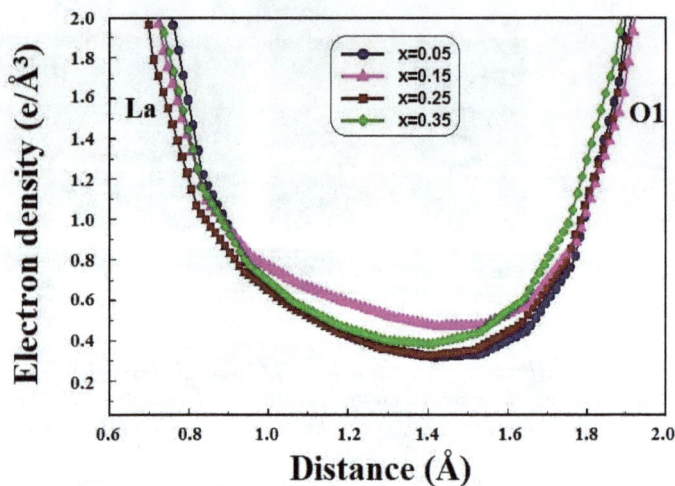

Figure 2.40 *One-dimensional charge density profiles along La-O1 bond in La$_{1-x}$Sr$_x$FeO$_3$ multiferroics*

Table 2.15 *Bond angles and bond lengths in the Fe-O sub-lattice of $La_{1-x}Sr_xFeO_3$ due to $<$ $FeO_6 >$ octahedral twist*

Samples	Bond angle (deg)		Bond length (Å)	
	Fe-O1-Fe	Fe-O2-Fe	Fe-O1	Fe-O2
x=0.05	158.0220	157.7021	1.9979	2.0703
x=0.10	161.8870	158.1390	1.9858	2.1305
x=0.15	162.8048	156.6920	1.9825	2.1179
x=0.20	164.2018	158.9486	1.9647	2.1693

Table 2.16 *Bond lengths and bond critical point charge densities of Fe-O1 and La-O1 bonds in $La_{1-x}Sr_xFeO_3$ multiferroics*

Samples	Fe-O1		La-O1	
	Bond length (Å)	Charge density (e/Å³)	Bond length(Å)	Charge density(e/Å³)
x=0.05	1.9979	0.8693	2.4351	1.0203
x=0.10	1.9858	0.8403	2.4988	0.8540
x=0.15	1.9825	0.7835	2.4844	0.7160

2.5 Optical characterization - UV-visible absorption spectra

The optical properties of the synthesized multiferroics have been analyzed by UV-visible absorption spectroscopy. The energy band gap has been estimated using UV- visible absorption data recorded in the wavelength range of 200 nm to 2000 nm. To estimate the energy band gap, Wood and Tauc [Wood and Tauc, 1972] method is used. In this section, the UV-visible absorption spectra and Tauc plot (plot of $(\alpha E)^2$ versus photon energy (E)) [Wood and Tauc, 1972] of all the synthesized multiferroics are presented. Theenergy band gap values of all the synthesized samples are given in Tables 2.17, 2.18, 2.19& 2.20.

2.5.1 Ce-substituted lanthanum orthoferrites - $La_{1-x}Ce_xFeO_3$

The UV-visible absorption spectra of the synthesized $La_{1-x}Ce_xFeO_3$ (x=0.00, 0.03, 0.06, 0.09 & 0.12) multiferroics are shown in the inset of Figure 2.41. The Tauc plot of $La_{1-x}Ce_xFeO_3$ has been drawn using the Wood and Tauc relation $\alpha h\upsilon = A(h\upsilon - E_g)^n$

[Wood and Tauc, 1972] and is presented in Figure 2.41. The energy band gap values of $La_{1-x}Ce_xFeO_3$ multiferroics are given in Table 2.17.

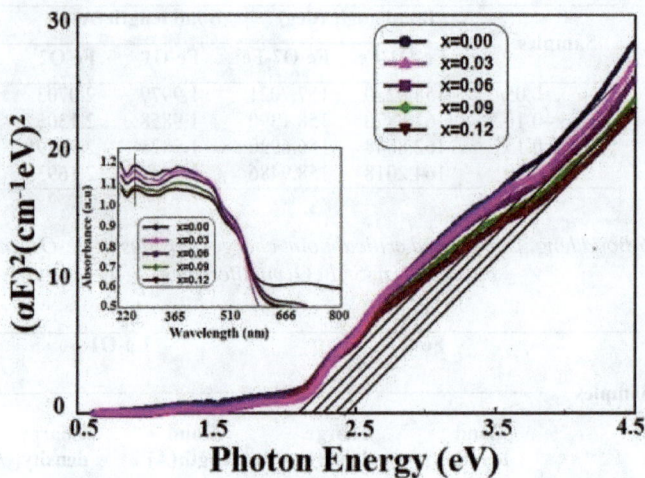

Figure 2.41 *Tauc plot of $La_{1-x}Ce_xFeO_3$ multiferroics (Inset: UV-visible absorption spectra of $La_{1-x}Ce_xFeO_3$)*

Table 2.17 *Energy band gap values of $La_{1-x}Ce_xFeO_3$ multiferroics*

Samples	Band gap (eV)
x=0.00	2.22(0.36)
x=0.03	2.23(0.56)
x=0.06	2.26(0.19)
x=0.09	2.34(0.13)
x=0.12	2.41(0.11)

2.5.2   Zn-substituted lanthanum orthoferrites - $La_{1-x}Zn_xFeO_3$

The UV-visible absorption spectra of the synthesized $La_{1-x}Zn_xFeO_3$ (x=0.00, 0.05, 0.15 & 0.25) multiferroics are shown in the inset of Figure 2.42. The Tauc plot of $La_{1-x}Zn_xFeO_3$

has been drawn using the Wood and Tauc method [Wood and Tauc, 1972] and is presented in Figure 2.42. The energy band gap values of $La_{1-x}Zn_xFeO_3$ are listed in Table 2.18.

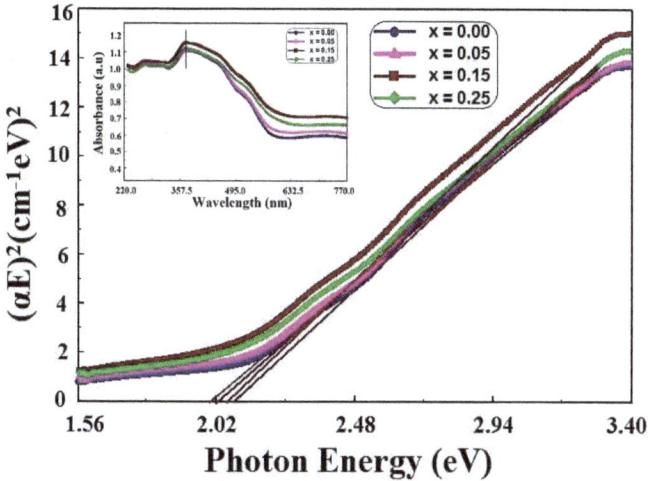

Figure 2.42 Tauc plot of $La_{1-x}Zn_xFeO_3$ multiferroics (Inset: UV-visible absorption spectra of $La_{1-x}Zn_xFeO_3$)

Table 2.18 Energy band gap values of $La_{1-x}Zn_xFeO_3$ multiferroics

Samples	Band gap (eV)
x=0.00	2.025(0.21)
x=0.05	2.048(0.2)
x=0.15	2.074(0.27)
x=0.25	2.097(0.13)

2.5.3 Al-substituted lanthanum orthoferrites - $La_{1-x}Al_xFeO_3$

The UV-visible absorption spectra of the synthesized $La_{1-x}Al_xFeO_3$ (x=0.05, 0.15 & 0.25) multiferroics are shown in the inset of Figure 2.43. The Tauc plot (photon energy

(E) versus $(\alpha E)^2$ plot) of $La_{1-x}Al_xFeO_3$ has been drawn to estimate the energy band gapand is presented in Figure 2.43. The estimated energy band gap values of $La_{1-x}Al_xFeO_3$ (x=0.05, 0.15 & 0.25) are tabulated in Table 2.19.

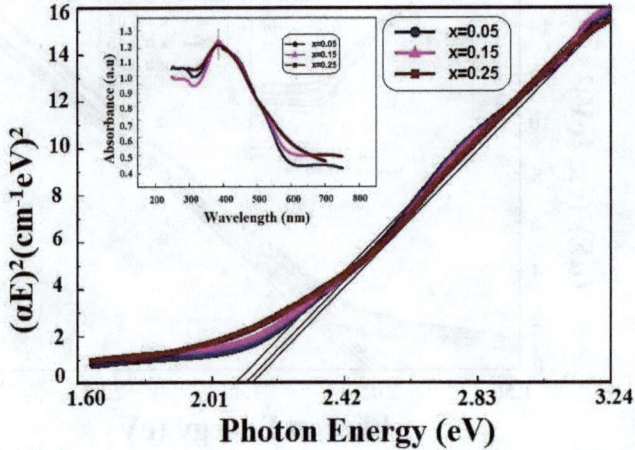

Figure 2.43 *Tauc plot of $La_{1-x}Al_xFeO_3$ multiferroics (Inset: UV-visible absorption spectra of $La_{1-x}Al_xFeO_3$)*

Table 2.19 *Energy band gap values of $La_{1-x}Al_xFeO_3$ multiferroics*

Samples	Band gap (eV)
x=0.05	2.129(0.2)
x=0.15	2.112(0.19)
x=0.25	2.095(0.24)

2.5.4 Sr-substituted lanthanum orthoferrites - $La_{1-x}Sr_xFeO_3$

The UV-visible absorption spectra of the synthesized $La_{1-x}Sr_xFeO_3$ (x=0.05, 0.10, 0.15 & 0.20) multiferroics are shown in the inset of Figure 2.44. To estimate the energy band gap for $La_{1-x}Sr_xFeO_3$, the Tauc plot has been drawn using the Wood and Tauc relation [Wood and Tauc, 1972] and is shown in Figure 2.44. The estimated direct energy band gap values of $La_{1-x}Sr_xFeO_3$ are presented in Table 2.20.

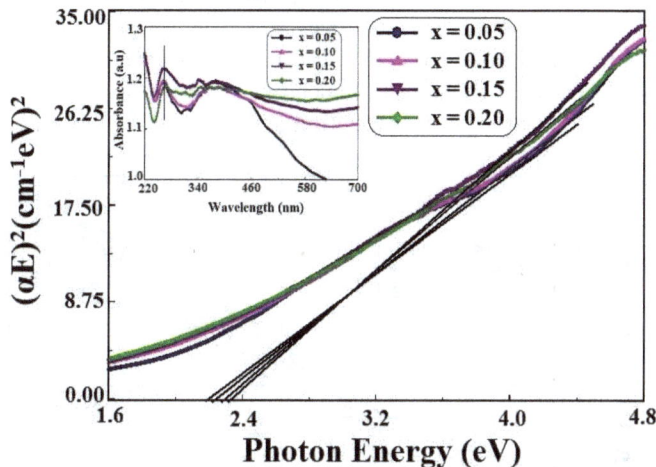

Figure 2.44 *Tauc plot of $La_{1-x}Sr_xFeO_3$ multiferroics (Inset: UV-visible absorption spectra of $La_{1-x}Sr_xFeO_3$)*

Table 2.20 *Energy band gap values of $La_{1-x}Sr_xFeO_3$ multiferroics*

Samples	Band gap (eV)
x=0.05	2.20(0.23)
x=0.10	2.24(0.34)
x=0.15	2.30(0.2)
x=0.20	2.36(0.39)

2.6 Magnetic characterization -Magnetic hysteresis

The magnetic characterization of all the synthesized multiferroics has been done using vibrating sample magnetometry. The magnetic hysteresis (M-H) curves of all the samples have been recorded at room temperature in the magnetic field range of $\pm$ 15000 G. In this section, the M-H curves of all the synthesized multiferroics are presented. The magnetic parameters such as saturation magnetization (M_s), remanent magnetization (M_r) and coercive field (H_C) of all the synthesized samples are tabulated and given in Tables 2.21, 2.22, 2.23 & 2.24.

2.6.1 Ce-substituted lanthanum orthoferrites - $La_{1-x}Ce_xFeO_3$

The room temperature magnetic hysteresis (M-H) curves of the synthesized $La_{1-x}Ce_xFeO_3$ (x=0.00, 0.03, 0.06, 0.09 & 0.12) multiferroics are shown in Figure 2.45 and the inset Figure 2.45 shows the enlarged M-H curves. The magnetic parameters such as saturation magnetization (M_s) and remanent magnetization (M_r), coercive field (H_C) and exchange bias field (H_{EB}) of $La_{1-x}Ce_xFeO_3$ are given in Table 2.21.

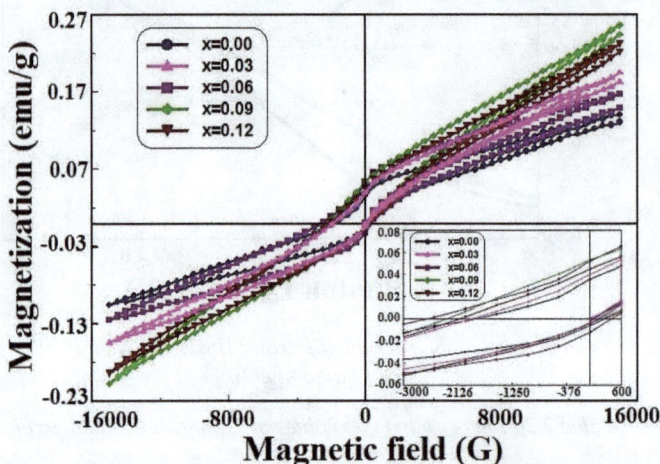

Figure 2.45 *Magnetic hysteresis (M-H) curves of $La_{1-x}Ce_xFeO_3$ multiferroics(Inset: Enlarged (M-H) curves)*

Table 2.21 *Magnetic parameters of $La_{1-x}Ce_xFeO_3$ multiferroics*

Samples	M_s (emu/g)	M_r (emu/g)	H_{C+} (G)	H_{C-} (G)	H_C (G)	H_{EB} (G)
x=0.00	0.141	0.036	123	1848	863	986
x=0.03	0.195	0.040	101	1911	905	1006
x=0.06	0.167	0.052	340	2639	1150	1490
x=0.09	0.256	0.053	165	2314	1075	1240
x=0.12	0.232	0.045	229	2081	926	1155

M_s - Saturation magnetizationM_r - Remanent magnetization

H_{C+}, H_{C-} - Intercepts of magnetization on the +ve and −ve side of the field axisH_C - Coercive field

H_{EB} - Exchange bias field.

2.6.2 Zn-substituted lanthanum orthoferrites - $La_{1-x}Zn_xFeO_3$

The room temperature magnetic hysteresis (M-H) curves of the synthesized $La_{1-x}Zn_xFeO_3$ (x=0.00, 0.05, 0.15 & 0.25) multiferroics are shown in Figure 2.46 and the inset Figure 2.46 shows the enlarged M-H curves. The magnetic parameters such as saturation magnetization (M_s) and remanent magnetization (M_r), coercive field (H_C) and exchange bias field (H_{EB}) of $La_{1-x}Zn_xFeO_3$ are listed in Table 2.22.

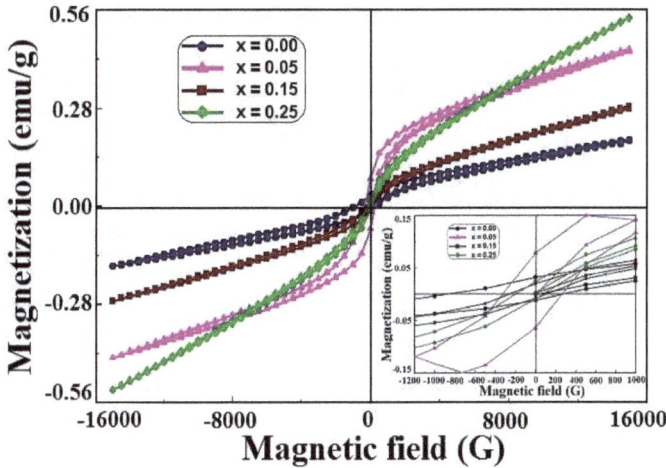

Figure 2.46 *Magnetic hysteresis (M-H) curves of $La_{1-x}Zn_xFeO_3$ multiferroics(Inset: Enlarged (M-H) curves)*

Table 2.22 *Magnetic parameters of $La_{1-x}Zn_xFeO_3$ multiferroics*

Samples	M_s (emu/g)	M_r (emu/g)	H_{C+} (G)	H_{C-} (G)	H_C (G)	H_{EB} (G)
x=0.00	0.18	0.022	275	865	295	570
x=0.05	0.44	0.072	267	325	29	296
x=0.15	0.28	0.014	115	261	73	188
x=0.25	0.53	0.017	80	198	59	139

M_s - Saturation magnetizationM_r - Remanent magnetization

H_{C+}, H_{C-} - Intercepts of magnetization on the +ve and −ve side of the field axisH_C - Coercive field

H_{EB} - Exchange bias field.

2.6.3   Al-substituted lanthanum orthoferrites - La$_{1-x}$Al$_x$FeO$_3$

The room temperature magnetic hysteresis (M-H) curves of the synthesized La$_{1-x}$Al$_x$FeO$_3$ (x=0.05, 0.15 & 0.25) multiferroics are shown in Figure 2.47. The inset Figure 2.47 shows the enlarged M-H curves of La$_{1-x}$Al$_x$FeO$_3$ (x=0.05 & x=0.15). The magnetic parameters such as saturation magnetization (M$_s$), remanent magnetization (M$_r$) and coercive field (H$_C$) of La$_{1-x}$Al$_x$FeO$_3$ are tabulated in Table 2.23.

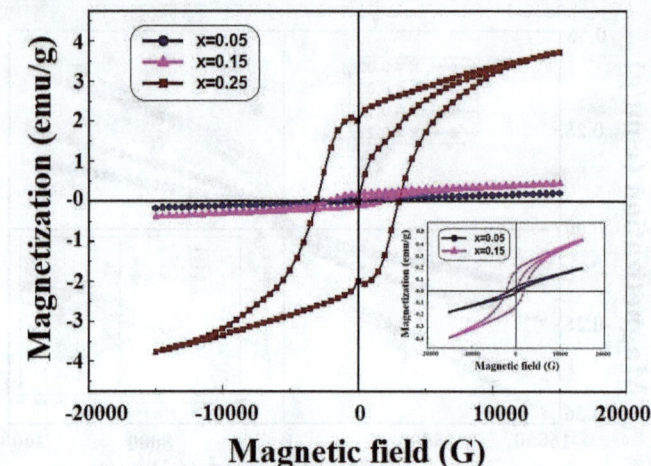

Figure 2.47 *Magnetic hysteresis (M-H) curves of La$_{1-x}$Al$_x$FeO$_3$ multiferroics(Inset: (M-H) curves of La$_{1-x}$Al$_x$FeO$_3$ (x=0.05 & x=0.15))*

Table 2.23 *Magnetic parameters of La$_{1-x}$Al$_x$FeO$_3$ multiferroics*

Samples	M$_s$ (emu/g)	M$_r$ (emu/g)	H$_C$ (G)
x=0.05	0.20	0.03	1044
x=0.15	0.44	0.17	1992
x=0.25	3.73	2.02	3031

M$_s$ - Saturation magnetizationM$_r$ - Remanent magnetizationH$_C$ - Coercive field.

2.6.4   Sr-substituted lanthanum orthoferrites - La$_{1-x}$Sr$_x$FeO$_3$

The room temperature magnetic hysteresis (M-H) curves of the synthesized La$_{1-x}$Sr$_x$FeO$_3$ (x=0.05, 0.10, 0.15 & 0.20) multiferroics are shown in Figure 2.48. The magnetic parameters such as saturation magnetization (M$_s$), remanent magnetization (M$_r$) and coercive field (H_C) of $La_{1-x}Sr_xFeO_3$ extracted from the hysteresis curves are given in Table 2.24.

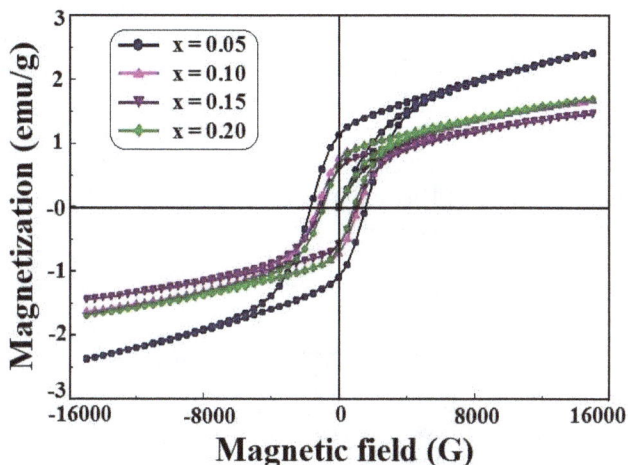

Figure 2.48 *Magnetic hysteresis (M-H) curves of $La_{1-x}Sr_xFeO_3$ multiferroics*

Table 2.24 *Magnetic parameters of $La_{1-x}Sr_xFeO_3$ multiferroics*

Samples	M_s (emu/g)	M_r (emu/g)	H_C (G)
x=0.05	2.4	1.12	1616
x=0.10	1.65	0.74	1210
x=0.15	1.46	0.59	1057
x=0.20	1.7	0.66	917

M_s - Saturation magnetizationM_r - Remanent magnetizationH_C - Coercive field

2.7 Dielectric characterization -Impedance analysis

The dielectric characterization for all the synthesized multiferroics has been done at room temperature using an impedance analyzer. The dielectric measurements (Impedance analysis) for all the samples have been carried out in the wide frequency range from 10 Hz to 5 MHz. In this section, the frequency-dependent dielectric constant, dielectric loss and ac conductivity plots of all the synthesized multiferroics are presented. The values of dielectric constant (ε'), dielectric loss ($\tan \delta$) and ac conductivity (σ_{ac}) of allthe synthesized samples are tabulated and given in Tables 2.25, 2.26, 2.27 & 2.28.

2.7.1 Ce-substituted lanthanum orthoferrites -La$_{1-x}$Ce$_x$FeO$_3$

Figure 2.49 shows the frequency (f) dependent dielectric constant (ε') plot of La$_{1-x}$Ce$_x$FeO$_3$ (x=0.00, 0.03, 0.06, 0.09 & 0.12) multiferroics and the inset Figure shows the frequency (f) dependent dielectric constant (ε') measured in the frequency range of 10 Hz to 5 MHz. The frequency (f) dependent dielectric loss (tan δ) plot of La$_{1-x}$Ce$_x$FeO$_3$ (x=0.00, 0.03, 0.06, 0.09 & 0.12) multiferroics is presented in Figure 2.50. Figure 2.51 illustrates the frequency (f) dependent ac conductivity (σ_{ac}) plot of La$_{1-x}$Ce$_x$FeO$_3$ (x=0.00, 0.03, 0.06 & 0.12) multiferroics and the inset Figure shows the (σ_{ac}-log f) plot of LaFeO$_3$. The values of dielectric constant (ε'), dielectric loss (tan δ) and ac conductivity (σ_{ac}) of La$_{1-x}$Ce$_x$FeO$_3$ are given in Table 2.25.

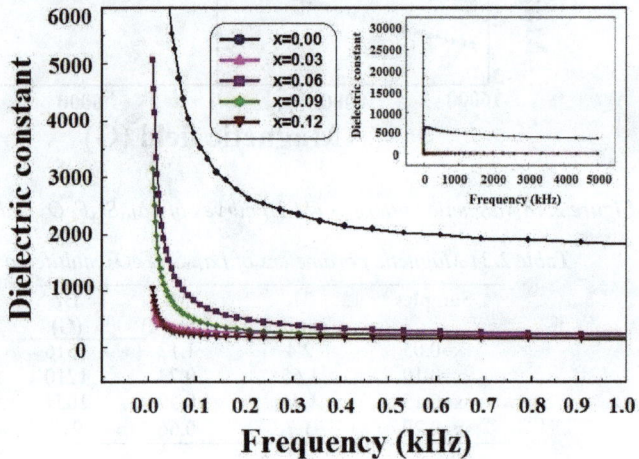

Figure 2.49 Frequency (f) dependent dielectric constant (ε') plot of La$_{1-x}$Ce$_x$FeO$_3$ (Inset: Frequency (f) dependent dielectric constant (ε') measured in the frequency range of 10 Hz to 5 MHz)

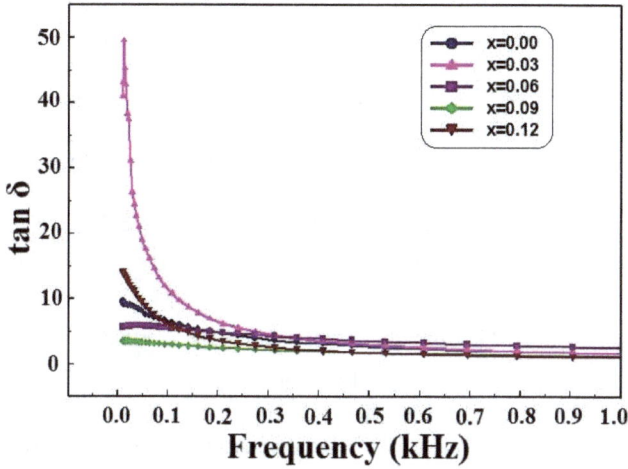

Figure 2.50 *Frequency (f) dependent dielectric loss (tan δ) plot of $La_{1-x}Ce_xFeO_3$*

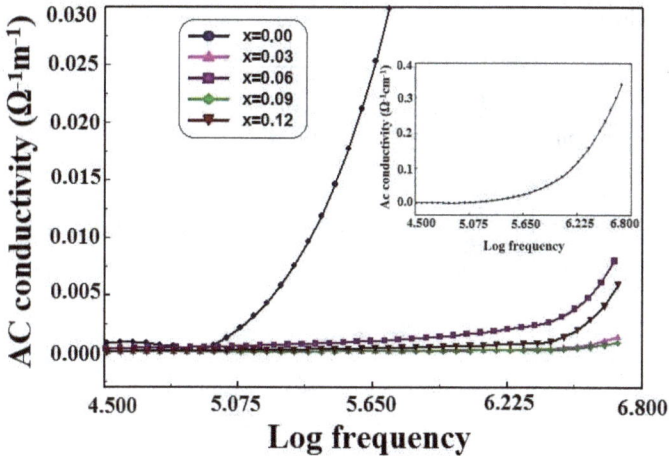

Figure 2.51 *Frequency (f) dependent ac conductivity (σ_{ac}) plot of $La_{1-x}Ce_xFeO_3$(Inset: (σ_{ac}-log f) plot of $LaFeO_3$)*

Table 2.25 *Dielectric parameters and ac conductivity of* $La_{1-x}Ce_xFeO_3$

Samples	[a]ε'	[a]$\tan\delta$	[b]σ_{ac} $(\Omega^{-1}m^{-1})$
x=0.00	27998	9.4	0.3389
x=0.03	811	41	0.0014
x=0.06	5070	5.7	0.0080
x=0.09	3158	3.5	0.0009
x=0.12	1032	13.9	0.0007

[a]ε' - Dielectric constant at 10 Hz [a]$\tan\delta$ - Dielectric loss factor at 10 Hz[b]σ_{ac} - AC conductivity at 5 MHz

2.7.2 Zn-substituted lanthanum orthoferrites - $La_{1-x}Zn_xFeO_3$

Figure 2.52 shows the frequency (f) dependent dielectric constant (ε') plot of $La_{1-x}Zn_xFeO_3$ (x=0.00, 0.05, 0.15 & 0.25) multiferroics and the inset Figure shows the frequency (f) dependence of dielectric constant (ε') measured in the frequency range of 10 Hz to 5 MHz. The frequency (f) dependent dielectric loss (tan δ) plot of $La_{1-x}Zn_xFeO_3$ (x=0.00, 0.05, 0.15 & 0.25) multiferroics is presented in Figure 2.53. Figure 2.54 illustrates the frequency (f) dependent ac conductivity (σ_{ac}) plot of $La_{1-x}Zn_xFeO_3$ (x=0.00, 0.05, 0.15 & 0.25) multiferroics. The values of dielectric constant (ε'), dielectric loss (tan δ) and ac conductivity (σ_{ac}) of $La_{1-x}Zn_xFeO_3$ are tabulated in Table 2.26.

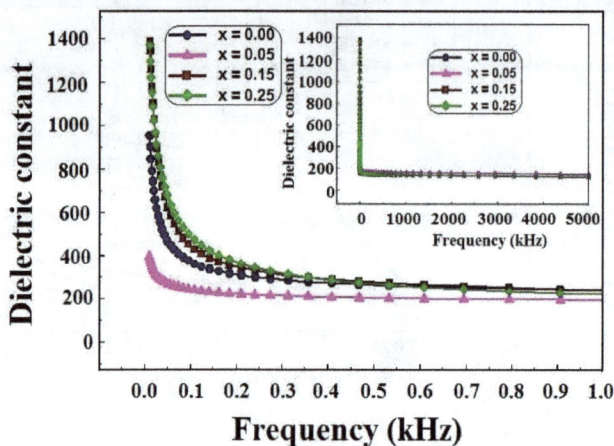

Figure 2.52 *Frequency (f) dependent dielectric constant (ε') plot of $La_{1-x}Zn_xFeO_3$ (Inset: Frequency (f) dependent dielectric constant (ε') measured in the frequency range of 10 Hz to 5 MHz)*

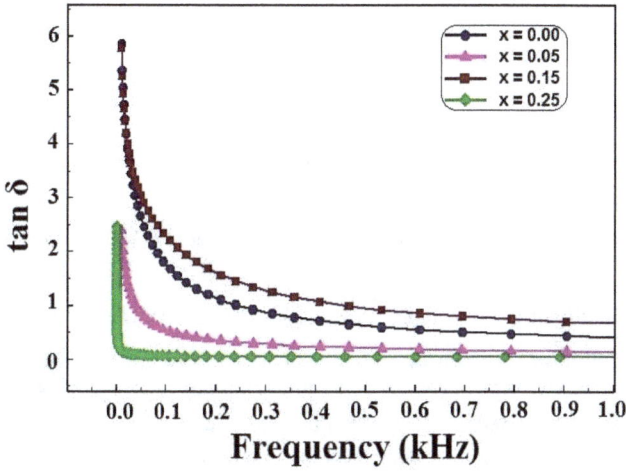

Figure 2.53 *Frequency (f) dependent dielectric loss (tan δ) plot of La$_{1-x}$Zn$_x$FeO$_3$*

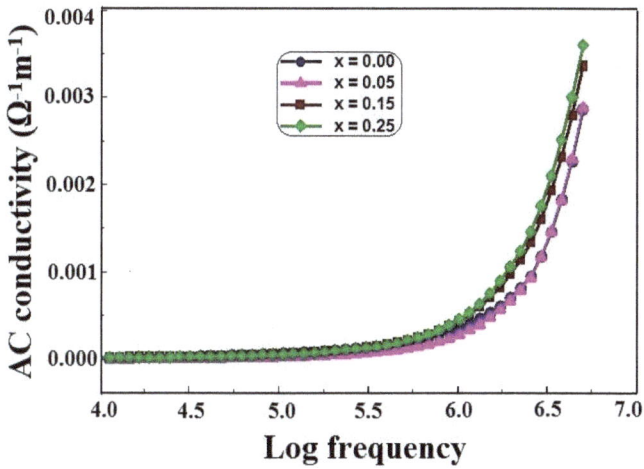

Figure 2.54 *Frequency dependent ac conductivity (σ$_{ac}$) plot of La$_{1-x}$Zn$_x$FeO$_3$*

Table 2.26 *Dielectric parameters and ac conductivity of La$_{1-x}$Zn$_x$FeO$_3$*

Samples	$^a\varepsilon'$	$^a tan\delta$	$^b\sigma_{ac}$ $(\Omega^{-1}m^{-1})$
x=0.00	955	5.9	0.0028
x=0.05	397	2.4	0.0029
x=0.15	1391	5.8	0.0034
x=0.25	1371	2.5	0.0036

$^a\varepsilon'$ - Dielectric constant at 10 Hz $^a tan\delta$ - Dielectric loss factor at 10 Hz $^b\sigma_{ac}$ - AC conductivity at 5 MHz

2.7.3   Al-substituted lanthanum orthoferrites - La$_{1-x}$Al$_x$FeO$_3$

Figure 2.55 shows the frequency (f) dependent dielectric constant (ε') plot of La$_{1-x}$Al$_x$FeO$_3$ (x=0.05, 0.15 & 0.25) multiferroics. The frequency (f) dependent dielectric loss (tan δ) plot of La$_{1-x}$Al$_x$FeO$_3$ (x=0.05, 0.15 & 0.25) multiferroics is illustrated in Figure 2.56. Figure 2.57 depicts the frequency (f) dependent ac conductivity (σ_{ac}) plot of La$_{1-x}$Al$_x$FeO$_3$ (x=0.05, 0.15 & 0.25) multiferroics. The values of dielectric constant (ε'), dielectric loss (tan δ) and ac conductivity (σ_{ac}) of La$_{1-x}$Al$_x$FeO$_3$ are given in Table 2.27.

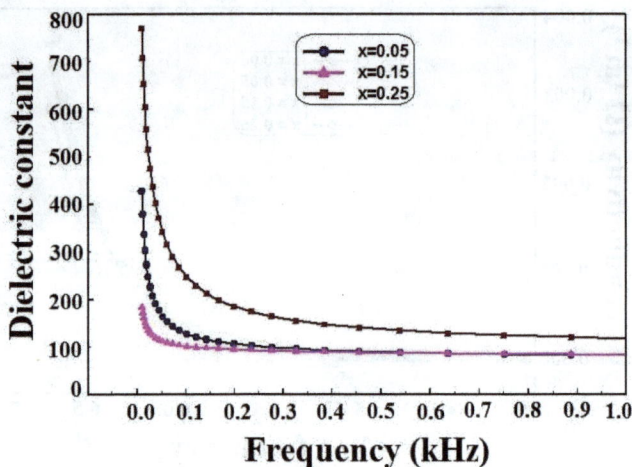

Figure 2.55 *Frequency (f) dependent dielectric constant (ε') plot of La$_{1-x}$Al$_x$FeO$_3$*

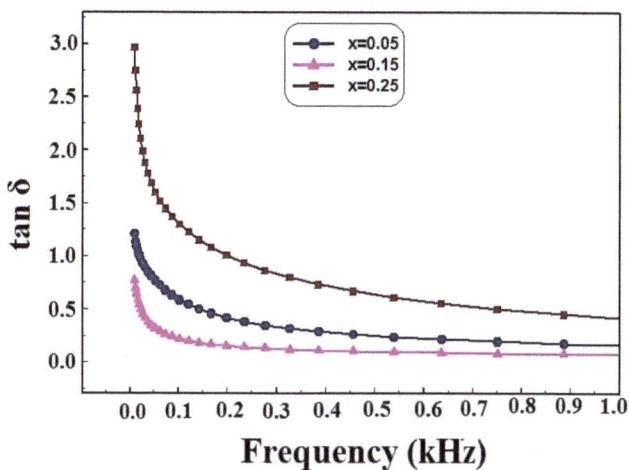

Figure 2.56 *Frequency (f) dependent dielectric loss (tan δ) plot of* $La_{1-x}Al_xFeO_3$

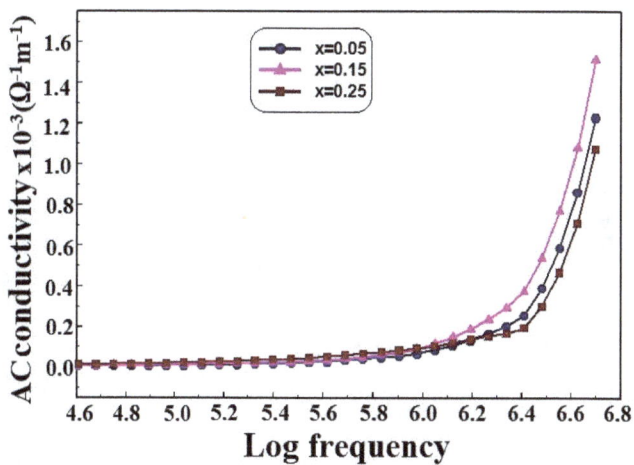

Figure 2.57 *Frequency dependent ac conductivity (σ_{ac}) plot of* $La_{1-x}Al_xFeO_3$

Table 2.27 *Dielectric parameters and ac conductivity of La$_{1-x}$Al$_x$FeO$_3$*

Samples	[a]ε'	[a]tanδ	[b]σ_{ac} x10^{-3}(Ω^{-1}m^{-1})
x=0.05	446	1.21	1.23
x=0.15	213	0.77	1.51
x=0.25	771	2.97	1.07

[a]ε' - Dielectric constant at 10 Hz [a]tanδ - Dielectric loss factor at 10 Hz[b]σ_{ac} - AC conductivity at 5 MHz.

2.7.4 Sr-substituted lanthanum orthoferrites - La$_{1-x}$Sr$_x$FeO$_3$

Figure 2.58 depicts the frequency (f) dependent dielectric constant (ε') plot of La$_{1-x}$Sr$_x$FeO$_3$ (x=0.05, 0.10, 0.15 & 0.20) multiferroics and the inset Figure shows the frequency (f) dependent dielectric constant (ε') plot of La$_{1-x}$Sr$_x$FeO$_3$ (x=0.10, 0.15 & 0.20). The frequency (f) dependent dielectric loss (tan δ) plot of La$_{1-x}$Sr$_x$FeO$_3$ (x=0.05, 0.10, 0.15 & 0.20) multiferroics is shown in Figure 2.59 and the inset Figure showstan δ-f plot of La$_{1-x}$Sr$_x$FeO$_3$ (x=0.05, 0.10 & 0.15). Figure 2.60 illustrates the frequency (f)dependent ac conductivity (σ_{ac}) plot of La$_{1-x}$Sr$_x$FeO$_3$ (x=0.05, 0.10, 0.15 & 0.20) multiferroics and the inset Figure shows the enlarged (σ_{ac}-log f) plot. The values of dielectric constant (ε'), dielectric loss (tan δ) and ac conductivity (σ_{ac}) of La$_{1-x}$Sr$_x$FeO$_3$ are presented in Table 2.28.

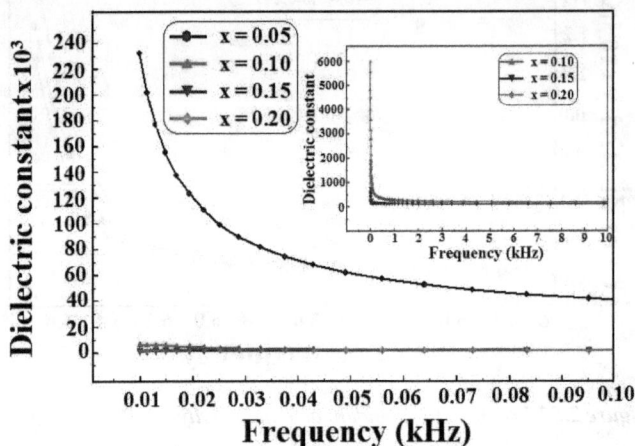

Figure 2.58 *Frequency (f) dependent dielectric constant (ε') plot of La$_{1-x}$Sr$_x$FeO$_3$ (Inset: (ε'-f) plot of La$_{1-x}$Sr$_x$FeO$_3$ (x=0.10, 0.15 & 0.20))*

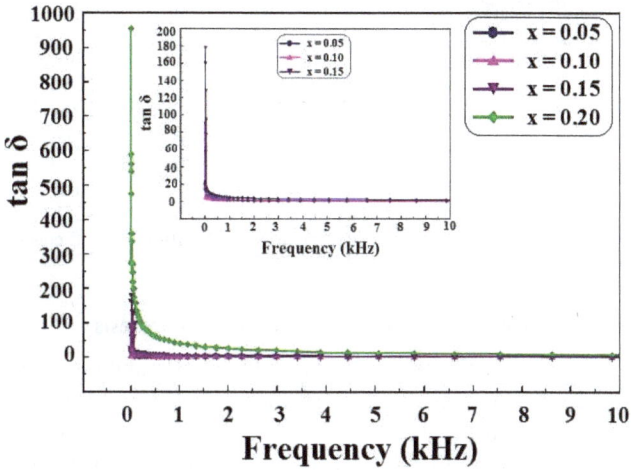

Figure 2.59 *Frequency (f) dependent dielectric loss (tan δ) plot of La$_{1-x}$Sr$_x$FeO$_3$ (Inset: (tan δ-f) plot ofLa$_{1-x}$Sr$_x$FeO$_3$ (x=0.10, 0.15 & 0.20))*

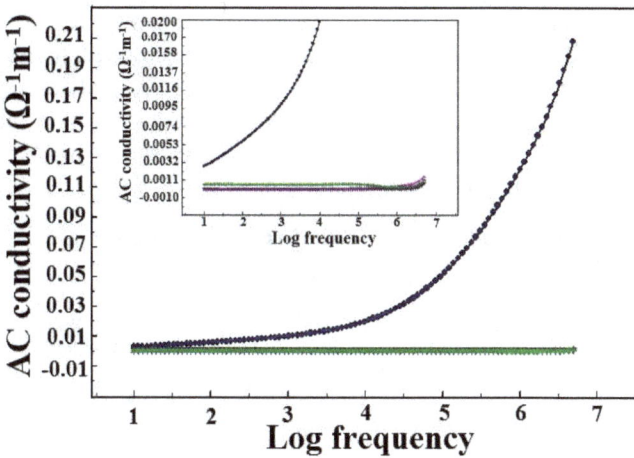

Figure 2.60 *Frequency dependence ac conductivity (σ$_{ac}$) plot of La$_{1-x}$Sr$_x$FeO$_3$ (Inset: Enlarged (σ$_{ac}$-log f)plot)*

Table 2.28 *Dielectric parameters and ac conductivity of* $La_{1-x}Sr_xFeO_3$

Samples	[a]ε'	[a]$\tan\delta$	[b]σ_{ac} $(\Omega^{-1}m^{-1})$
x=0.05	232415	21.5	0.2081
x=0.10	5979	4.95	0.0015
x=0.15	736	178	0.0006
x=0.20	1911	955	0.0011

[a]ε' - Dielectric constant at 10 Hz [a]$\tan\delta$ - Dielectric loss factor at 10 Hz[b]σ_{ac} - AC conductivity at 5 MHz.

2.8 Ferroelectric characterization -Ferroelectric hysteresis

The ferroelectric characterization of all the synthesized multiferroics has been done using the Ferroelectric loop tracer. The ferroelectric hysteresis (polarization (P) versus electric field (E)) curves or loops of the synthesized $La_{1-x}Ce_xFeO_3$, $La_{1-x}Zn_xFeO_3$ and $La_{1-x}Al_xFeO_3$ multiferroics have been recorded at room temperature at different applied electric fields and frequencies. As $La_{1-x}Sr_xFeO_3$ (x=0.05, 0.10, 0.15 & 0.20) multiferroics do not respond to the different applied electric fields their ferroelectric hysteresis loops could not be observed. The reason for this is explained as follows: The substitution of Sr in $LaFeO_3$ leads to the oxidation of Fe^{3+} ions into Fe^{4+} ions to maintain charge neutrality in the system. The presence of Fe ions with mixed-valence states makes the synthesized $La_{1-x}Sr_xFeO_3$ less insulating and so their ferroelectric loops could not be observed. In this section, the P-E hysteresis loops of the synthesized $La_{1-x}Ce_xFeO_3$, $La_{1-x}Zn_xFeO_3$ and $La_{1-x}Al_xFeO_3$ multiferroics are presented. The ferroelectric parameters such as maximum electric polarization (P_m), remanent polarization (P_r) and electric coercive field (E_C) of the samples are tabulated and given in Tables 2.29, 2.30 & 2.31.

2.8.1 Ce-substituted lanthanum orthoferrites - $La_{1-x}Ce_xFeO_3$

Figures 2.61(a)-(e) illustrate the P-E hysteresis loops of $La_{1-x}Ce_xFeO_3$ (x=0.00, 0.03, 0.06, 0.09 & 0.12) multiferroics. The ferroelectric parameters such as maximum electric polarization (P_m), remanent polarization (P_r) and electric coercive field (E_C)extracted from the hysteresis loops are given in Table 2.29.

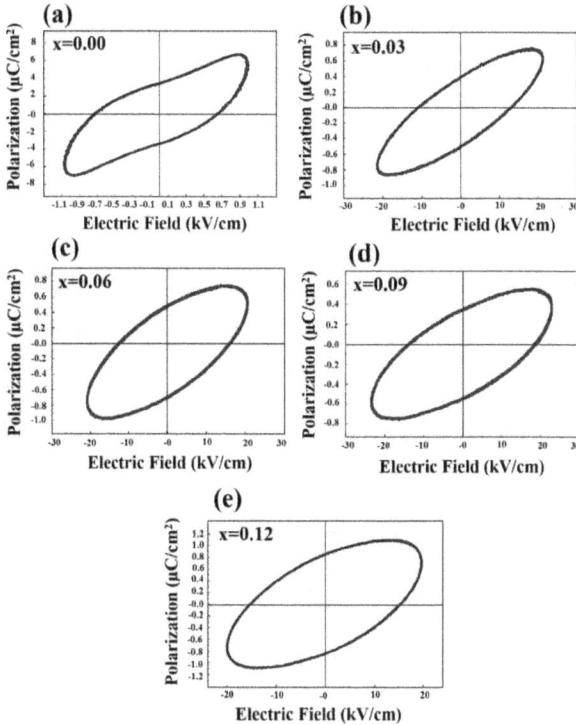

Figure 2.61 *P-E hysteresis loops of La$_{1-x}$Ce$_x$FeO$_3$ multiferroics,* *(a)* *x=0.00 (observed at field 1 kV/cm, frequency 50 Hz) and* *(b)* *x=0.03* *(c)* *x=0.06* *(d)* *x=0.09 &* *(e)* *x=0.12 (observed at field 20 kV/cm, frequency 20 Hz)*

Table 2.29 *Ferroelectric parameters of La$_{1-x}$Ce$_x$FeO$_3$ multiferroics*

Samples	Applied field (kV/cm)	Frequency (Hz)	P$_m$ (μC/cm^2)	P$_r$ (μC/cm^2)	E$_C$ (kV/cm)
x=0.00	1	50	6.83	3.42	0.7
x=0.03	20	20	0.81	0.45	12
x=0.06	20	20	0.86	0.59	14.3
x=0.09	20	20	0.65	0.44	16.3
x=0.12	20	20	1.09	0.83	15.3

P$_m$ - Maximum electric polarization P$_r$ - Remanent polarization, E$_C$ - Electric coercive field

2.8.2 Zn-substituted lanthanum orthoferrites -$La_{1-x}Zn_xFeO_3$

The P-E hysteresis loops of $La_{1-x}Zn_xFeO_3$ (x=0.00, 0.05, 0.15 & 0.25) multiferroics are shown in Figures 2.62 (a)-(d). The ferroelectric parameters such as maximum electric polarization (P_m), remanent polarization (P_r) and electric coercive field (E_C) extracted from the P-E loops are listed in Table 2.30.

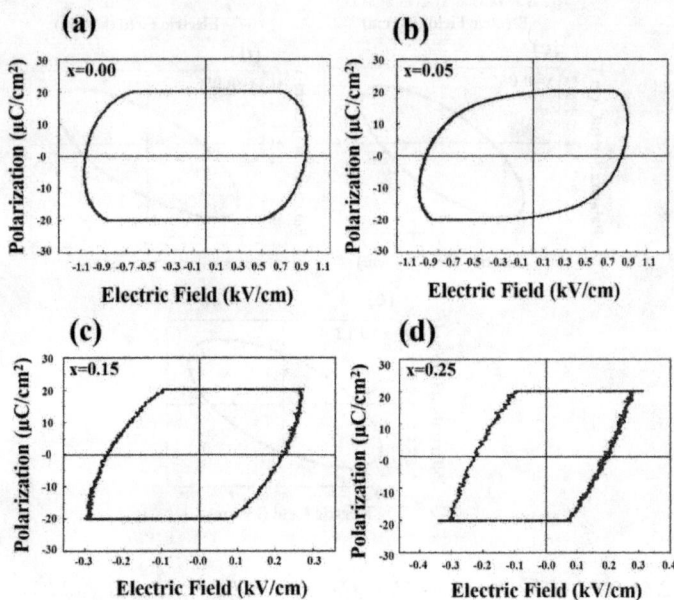

Figure 2.62 *P-E hysteresis loops of $La_{1-x}Zn_xFeO_3$ multiferroics*

Table 2.30 *Ferroelectric parameters of $La_{1-x}Zn_xFeO_3$ multiferroics*

Samples	P_m ($\mu C/cm^2$)	P_r ($\mu C/cm^2$)	E_C (kV/cm)
x=0.00	20.01	20.01	0.44
x=0.05	20	19.26	0.94
x=0.15	20.11	20.11	0.23
x=0.25	20.12	20.12	0.21

P_m - Maximum electric polarization P_r - Remanent polarization, E_C - Electric coercive field

2.8.3   Al-substituted lanthanum orthoferrites - La$_{1-x}$Al$_x$FeO$_3$

Figures 2.63 (a)-(c) show the P-E hysteresis loops of La$_{1-x}$Al$_x$FeO$_3$ (x=0.05, 0.15 & 0.25) multiferroics. The ferroelectric parameters such as maximum electric polarization(P_m), remanent polarization (P_r) and electric coercive field (E_C) obtained from the ferroelectric hysteresis loops are summarized in Table 2.31.

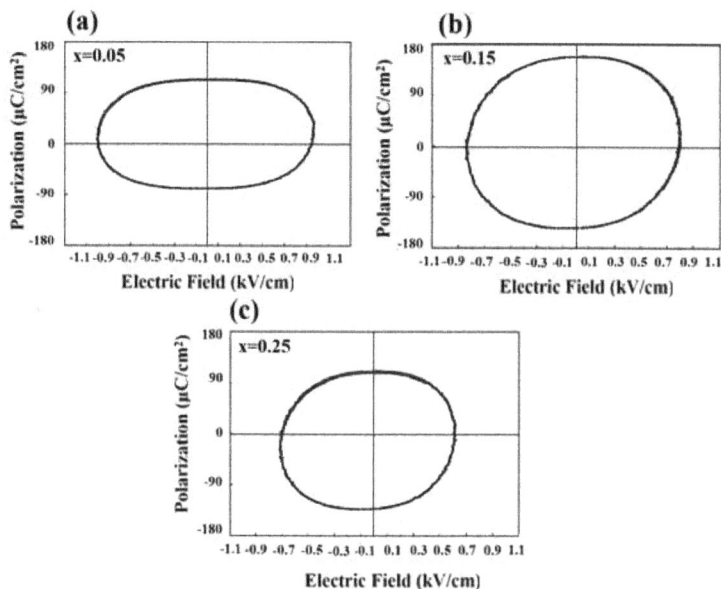

Figure 2.63 *P-E hysteresis loops of La$_{1-x}$Al$_x$FeO$_3$ multiferroics*

Table 2.31 *Ferroelectric parameters of La$_{1-x}$Al$_x$FeO$_3$ multiferroics*

Samples	P_m ($\mu C/cm^2$)	P_r ($\mu C/cm^2$)	E_C (kV/cm)
x=0.05	96.53	96.50	0.97
x=0.15	151.4	151.3	0.81
x=0.25	122	120	0.65

P_m - Maximum electric polarizationP_r - Remanent polarization, E_C - Electric coercive field

References

[1] Charles Robert M., Saravanan R., Powder Technol., 197, 159 (2010).
https://doi.org/10.1016/j.powtec.2009.09.009

[2] Collins D. M, Nature, 298, 49 (1982). https://doi.org/10.1038/298049a0

[3] Holland T.J.B., Redfern S.A.T., Mineral Mag., 61, 65 (1997).
https://doi.org/10.1180/minmag.1997.061.404.07

[4] Izumi F., Dilanien R.A., Recent Research Developments in Physics, Vol.3, Part II,
Transworld Research Network. Trivandrum, (2002).

[5] Kannan Y.B., Saravanan R., Srinivasan N., Ismail I., J. Magn.Magn. Mater., 423, 217
(2017). https://doi.org/10.1016/j.jmmm.2016.09.038

[6] Meenakshi S.V., Saravanan R., Srinivasan N., Saravanan O.V., Dhayanithi D.,
Giridharan N.V., J. Electron. Mater., 49, 7349 (2020).
https://doi.org/10.1007/s11664-020-08481-4

[7] Momma K., Izumi F., VESTA: a three-dimensional visualization system for
electronic and structural analysis. J. Appl. Crystallogr., 41, 653 (2008).
https://doi.org/10.1107/S0021889808012016

[8] Petricek V., Dusek M., Palatinus L., Kristallogr Z, Crystallographic Computing
System JANA2006: General features., 229, 345 (2014).
https://doi.org/10.1515/zkri-2014-1737

[9] Rietveld H.M., J. Appl. Crystallogr., 2, 65 (1969).
https://doi.org/10.1107/S0021889869006558

[10] Saravanakumar S., Escobedo-Morales, Pal U., Aranda R.J., Saravanan R., J. Mater.
Sci.,49, 5529 (2014). https://doi.org/10.1007/s10853-014-8242-z

[11] Saravanan R., Thirumalaisamy T.K., Kajitani T., Phys. Status Solidi A., 208, 2643
(2011). https://doi.org/10.1002/pssa.201127184

[12] Sasikumar S., Saravanan R., J.Electron. Mater., 46, 4187 (2017).
https://doi.org/10.1007/s11664-017-5349-4

[13] Wood D.L., Tauc J., Phys Rev B., 5, 3144 (1972).
https://doi.org/10.1103/PhysRevB.5.3144

Multiferroic Materials
Materials Research Foundations **140** (2023)

Materials Research Forum LLC
https://doi.org/10.21741/9781644902271

Chapter 3

Analysis of Results

Abstract

The experimental results obtained from various characterization methods such as powder X-ray diffraction (PXRD), scanning electron microscopy (SEM),energy dispersive X-ray spectroscopy (EDS), UV-visible absorption spectroscopy (UV- Vis), vibrating sample magnetometry (VSM), dielectric and ferroelectric characterizations, for all the synthesized multiferroics have been analyzed in detail. Furthermore, the charge density distribution and interatomic chemical bonding behavior obtained by the Maximum Entropy Method (MEM) [Collins, 1982] for all the multiferroics have been analyzed in detail. The experimentally observed optical, magnetic, dielectric and ferroelectric properties have also been investigated through charge densities.

Keywords

X-Ray Diffraction (PXRD), Scanning Electron Microscopy (SEM),Energy Dispersive X-Ray Spectroscopy (EDS), UV-Visible Absorption Spectroscopy (UV- Vis), Vibrating Sample Magnetometry (VSM), Dielectric and Ferroelectric Characterizations

3.1 Introduction

In this chapter, the experimental results obtained from various characterization methods such as powder X-ray diffraction (PXRD), scanning electron microscopy (SEM),energy dispersive X-ray spectroscopy (EDS), UV-visible absorption spectroscopy (UV- Vis), vibrating sample magnetometry (VSM), dielectric and ferroelectric characterizations, for all the synthesized multiferroics have been analyzed in detail. Furthermore, the charge density distribution and interatomic chemical bonding behavior obtained by the Maximum Entropy Method (MEM) [Collins, 1982] for all the multiferroics have been analyzed in detail. The experimentally observed optical, magnetic, dielectric and ferroelectric properties have also been investigated through charge densities.

3.2 Synthesis of lanthanum orthoferrite-type multiferroics

In this work, four series of lanthanum orthoferrite (LFO)-type multiferroics have been synthesized by conventional high temperature solid state reaction method [Saravanan and Mangaiyarkkarasi, 2016, Thenmozhi et al., 2016, Sasikumar et al., 2017]. The calcination temperatures, sintering temperatures and their corresponding time duration, grinding duration followed to synthesis lanthanum orthoferrite-type multiferroics are tabulated for comparison and are presented in Table 3.1.

Table 3.1 *Calcination, sintering temperatures and grinding duration followed to synthesislanthanum orthoferrite (LFO)-type multiferroics*

| Samples | Calcination | | | I Sintering | | | II Sintering | |
	Temp.	Durn.	Grindig Duration	Temp.	Durn.	Regriding Duration	Temp.	Durn.
	(° C)	(h)	(h)	(° C)	(h)	(h)	(° C)	(h)
LCFO	1000	10	10	1300	16	6	1300	16
LZFO	1000	10	10	1200	12	10	1300	16
LAFO	1000	10	10	1000	10	10	1300	16
LSFO	1000	10	10	1300	16	6	1300	16

LCFO - $La_{1-x}Ce_xFeO_3$;
LZFO - $La_1-Zn_xFeO_3$;
LAFO - $La_{1-x}Al_xFeO_3$;
LSFO - $La_{1-x}Sr_xFeO_3$
Temp.- Temperature; Durn. - Duration

3.3 Structural analysis using powder X-ray diffraction

The structural characterization of all the synthesized multiferroics has been done using powder X-ray diffraction (PXRD). To investigate the structural changes due to the substitution of various elements such as Ce, Zn, Al and Sr at the lattice site of La in $LaFeO_3$, all the experimental XRD data sets have been subjected to Rietveld [Rietveld, 1969] refinement through the software program JANA 2006 [Patricek et al., 2014]. The detailed discussion of experimentally observed XRD profiles and the powder profile refinement is given in this section.

3.3.1 Ce-substituted lanthanum orthoferrites - $La_{1-x}Ce_xFeO_3$

The powder X-ray diffraction (XRD) patterns of the synthesized ($La_{1-x}Ce_xFeO_3$) (x=0.00, 0.03, 0.06, 0.09 & 0.12) multiferroics (Figure 2.1 (a)) confirm that all the synthesized products are in the orthorhombic phase with space group *Pnma* (Space group No: 62) [Aroyo, 2016]. In this work, for XRD data analysis, we have used Joint Committee for Powder Diffraction Standards (JCPDS) formats and the references corresponding to the JCPDS formats instead of International centre for diffraction data (ICDD) formats. All the observed X-ray diffraction peaks of $La_{1-x}Ce_xFeO_3$ multiferroics match well JCPDS data corresponding to $LaFeO_3$ [JCPDS PDF# 37-1493]. The XRD patterns indicate that there is an additional phase only for x=0.09 and 0.12 and for all other concentrations x=0.00, 0.03 and 0.06 only one phase appears. The additional phase is identified as CeO_2 from JCPDS database [JCPDS PDF# 43-1002]. The structure of the additional phase CeO_2 belongs to a cubic system with space group *Fm-3m* (Space group No: 225) [Aroyo, 2016]. From the XRD data, it is observed that for x=0.03, 0.06 and 0.09, the peaks are shifted towards the higher Bragg's angles from the peaks for x=0.00. This observation reveals that for Ce concentration 3% - 9%, the orthorhombic lattice of $LaFeO_3$ has shrunk due to the substitution of the cation with smaller ionic radius Ce^{4+} (r_{Ce}=0.87 Å) [Shannon, 1976] at the site of the cation with a larger ionic radius La^{3+} (r_{La}=1.03 Å) [Shannon, 1976]. But for the highest Ce concentration x=0.12, the Bragg peaks are shifted towards the lower Bragg's angles from the peaks for x=0.09. The reason for this left shift can be explained as follows: Cerium can exist in the oxidation states Ce^{3+} and Ce^{4+}, and during sintering some Ce^{4+} ions (r_{Ce}=0.87 Å) may be reduced into Ce^{3+} (r_{Ce}=1.01 Å) [Shannon, 1976] with high ionic radius. The existence of Ce in Ce^{3+} ions (r_{Ce}=1.01 Å) [Shannon, 1976] oxidation state in the host lattice is the cause for the shift of XRD peaks towards the lower 2θ side at this concentration (x=0.12). The enlarged views of the XRD peak of (121) crystallographic plane of $La_{1-x}Ce_xFeO_3$ (Figure 2.1(b)) show the shift of XRD peaks clearly. This XRD peak shift that arises due to Ce substitution in $LaFeO_3$ is evident for the change in structural parameters of $La_{1-x}Ce_xFeO_3$. The crystal structures of $La_{1-x}Ce_xFeO_3$ (x=0.00, 0.03, 0.06, 0.09 & 0.12) multiferroics have been refined using Rietveld refinement [Rietveld, 1969] technique by employing JANA 2006 [Patricek et al., 2014] software program (Figures 2.2 (a)-(e)). The profile refinement was performed by considering $LaFeO_3$ to be an orthorhombic crystal system with the space group *Pnma* (Space group No: 62) [Aroyo, 2016] and 4 molecules per unit cell. The atom position was fixed at Wyckoff position 4c (0.0295, 0.25, 0.9951) for La/Ce, 4b (0.5, 0, 0) for Fe, 4c (0.488, 0.25, 0.0707) for apex atom O1 and 8d (0.2831, 0.0387, 0.7168) for planar atom O2 [Wyckoff, 1963]. The structure of the additional phase CeO_2 was refined by considering the cubic symmetry with *Fm-3m* space group (Space group No: 225) [Aroyo, 2016] and 4 molecules per unit cell. The initial atomic positional coordinates were taken as (0, 0, 0) for Ce atom and (0.25, 0.25, 0.25) for O atom [Wyckoff, 1963]. The cell parameters of $La_{1-x}Ce_xFeO_3$ (x=0.00, 0.03, 0.06, 0.09 & 0.12) multiferroics were refined by the unit cell refinement software program [Holland and Redfern, 1997]. The refined cell parameters obtained from the unit cell refinement method [Holland and Redfern, 1997] and the profile fitted parameters extracted from the Rietveld refinement method [Rietveld, 1969] (Table 2.1) indicate that up to x=0.09, the orthorhombic unit cell parameters and also the unit cell volume tends to decrease with the increase in Ce concentration. This is due to the replacement of La^{3+} ion with larger ionic radius (r_{La} =1.03 Å) [Shannon, 1976] by Ce^{4+} ion with smaller ionic radius (r_{Ce}=0.87 Å) [Shannon, 1976]. The values of profile fitted parameters such as reliability index for profile (R_P), reliability index for observed structure factors (R_{obs}) and goodness of fit (GOF) (Table 2.1) indicate the reliable fitting between the observed and calculated XRD profiles for all the synthesized multiferroics.

3.3.2 Zn-substituted lanthanum orthoferrites - $La_{1-x}Zn_xFeO_3$

The observed X-ray diffraction patterns of $La_{1-x}Zn_xFeO_3$ (x=0.00, 0.05, 0.15 & 0.25) multiferroics (Figure 2.3 (a)) confirm that all samples crystallize in distorted orthorhombic structure with space group *Pnma* (Space group No: 62) [Aroyo, 2016] and the diffraction peaks match well with the standard data corresponding to $LaFeO_3$ [JCPDS PDF# 37-1493]. The XRD patterns indicate that the pristine sample is monophasic and noadditional phase is identified in this system, whereas in the XRD patterns of x=0.05, 0.15 and 0.25, additional peaks are identified, and they match well with the standard data corresponding to Fe_3O_4 [JCPDS PDF# 11-0614]. The crystal structure of the additional phase Fe_3O_4 is found to be a cubic system of space group *Fd-3m* (Space group No: 227) [Aroyo, 2016], and the peaks related to it are marked as '*' (Figure 2.3(a)). The XRD patterns (Figure 2.3 (a)) indicate that most of the Bragg peaks corresponding to x=0.05 are shifted towards the higher 2θ angles with respect to x=0.00, whereas for x=0.15 and 0.25, most of the peaks are shifted slightly towards the lower 2θ angles with respect to x=0.05 concentration. The observed peak shift is due to the charge compensation effect in the host matrix, which can be explained as follows: The substitution of divalent ion Zn^{2+} in the place of trivalent ion La^{3+} in $LaFeO_3$ causes charge imbalance and disturbs the charge neutrality of the system. To maintain the charge neutrality in the host matrix, cation Fe^{3+} (0.645 Å) [Shannon, 1976] reduces to Fe^{2+} (0.78 Å) [Shannon, 1976] with the creation of oxygen vacancies, which results in the formation of additional phase Fe_3O_4 [Manzoor and Husain, 2018]. Thus, the existence of Fe^{2+} ions in the host matrix is the cause for the shifting of XRD peaks towards the lower 2θ angles. Likewise, the reason forthe different tendency of peak shift (right shift) for x=0.05 compared with x=0.15 and

0.25 is due to the presence of less number of Fe^{2+} ions in this concentration. The enlarged XRD peak corresponding to (121) crystallographic plane (Figure 2.3(b)) of the samples clearly explains the shifting of Bragg peaks. Thus, the substitution of Zn in $LaFeO_3$ creates the charge compensation of Fe^{3+} ions in the host crystal lattice (wherein Fe^{3+} ions were reduced to Fe^{2+} ions), which results in the decrease in the number of Fe^{3+} ions. In addition, Zn substitution also decreases the particle size and as a result, increases the uncompensated spins of Fe^{3+}. Thus, the observed peak shift reflects that the Zn atom is substituted in the preferred sites of the host matrix.

The experimental X-ray profiles of $La_{1-x}Zn_xFeO_3$ (x=0.00, 0.05, 0.15 & 0.25) multiferroics were refined by considering an orthorhombic system with the space groupof $Pnma$ (Space group No: 62) [Aroyo, 2016] and 4 molecules per unit cell using Rietveld refinement [Rietveld, 1969] technique, which employs the software program JANA2006 [Patricek et al., 2014] (Figures 2.4 (a)-(d)). The positional coordinates of the atom La/Zn was fixed at Wyckoff position 4c (0.0295, 0.25, 0.9951), atom Fe was at 4b (0.5, 0, 0), apex atom O1 was at 4c (0.488, 0.25, 0.0707) and planar atom O2 was at 8d (0.2831, 0.0387, 0.7168) [Wyckoff, 1963]. The refinement for the additional phase Fe_3O_4 was performed with cubic symmetry, Fd-$3m$ space group (Space group No: 227) [Aroyo, 2016], and 8 molecules per unit cell. The positional coordinates for the octahedral site atom Fe1 was fixed at 16d (0.5, 0.5, 0.5), the tetrahedral site atom Fe2 was at 8a (0.125, 0.125, 0.125) and atom O was at 32e (0.25, 0.25, 0.25) [Glazyrin et al., 2012]. The structural parameters of the synthesized $La_{1-x}Zn_xFeO_3$ (x=0.00, 0.05, 0.15 & 0.25) multiferroics were refined using the unit cell refinement software program [Holland and Redfern, 1997]. The structural parameters obtained from the unit cell refinement method [Holland and Redfern, 1997] and the profile fitted parameters extracted from the Rietveld refinement method [Rietveld, 1969] (Table 2.2) show that the unit cell parameters and also the unit cell volume non-monotonically decrease with the increase in the substitution of Zn, due to the reduction of Fe^{3+} ions into Fe^{2+} [Manzoor and Husain, 2018] to balance the charges in the host matrix. The values of reliability indices and goodness of fit indicate a perfect fit between the observed and calculated XRD profiles (Table 2.2).

3.2.3 Al-substituted lanthanum orthoferrites - $La_{1-x}Al_xFeO_3$

The raw XRD patterns of $La_{1-x}Al_xFeO_3$ (x=0.05, 0.15 & 0.25) multiferroics (Figure 2.5(a)) show that the structure of the synthesized $La_{1-x}Al_xFeO_3$ belongs to the perovskite-based orthorhombic structure with space group $Pnma$ (Space group No: 62) [Aroyo, 2016]. The XRD peaks match well with JCPDS data of $LaFeO_3$ [JCPDS PDF# 37-1493] and they reveal that all samples are monophasic. The X-ray diffraction peaks shift towards the higher 2θ angles with the increase in Al concentration in the host lattice. The enlarged XRD peak corresponding to (121) crystallographic plane (Figure 2.5(b)) clearly depicts that the peaks shift towards the higher 2θ angles with the increase in the Alconcentration. The shift of diffraction peaks towards the higher 2θ angles confirms that the orthorhombic unit cells of $La_{1-x}Al_xFeO_3$ reduced in size due to the substitution of Al^{3+} ions (r_{Al}=0.535 Å) [Shannon, 1976] having smaller ionic radii at the lattice site of the larger La^{3+} (r_{La}=1.03 Å) [Shannon, 1976] ions.

The crystal structure refinement for the synthesized $La_{1-x}Al_xFeO_3$ (x=0.05, 0.15 & 0.25) multiferroics was performed using Rietveld refinement [Rietveld, 1969] technique by considering orthorhombic perovskite structure with *Pnma* space group (Space group No: 62) [Aroyo, 2016] and four molecules per unit cell. The Rietveld refinement [Rietveld, 1969] technique uses the least square approach to minimize the difference between the observed XRD profile and theoretically constructed profile, which results in an accurate structural model. Figures 2.6 (a)-(c) illustrate the fitted Rietveld [Rietveld, 1969] XRD profiles of $La_{1-x}Al_xFeO_3$ (x=0.05, 0.15 & 0.25) multiferroics. The structural and profile fitted parameters extracted from the Rietveld refinement method [Rietveld, 1969] (Table 2.3) indicate that the cell parameters and unit cell volume decrease with the increase inthe substitution of Al. The decrease in cell parameters and cell volume is due to the substitution of smaller Al^{3+} (r_{Al}=0.535 Å) [Shannon, 1976] ions at the lattice site of larger La^{3+} (r_{La}=1.03 Å) [Shannon, 1976] ions. The profile fitted parameters R_p, R_{obs}, and goodness of fit (GOF)) extracted from Rietveld refinement [Rietveld, 1969] suggest that the theoretically constructed XRD profiles match well with the experimentally observed one, which confirms the orthorhombic structure of $La_{1-x}Al_xFeO_3$ multiferroics.

3.2.4 Sr-substituted lanthanum orthoferrites - $La_{1-x}Sr_xFeO_3$

The experimental XRD patterns of $La_{1-x}Sr_xFeO_3$ (x=0.05, 0.10, 0.15 & 0.20) multiferroics (Figure 2.7 (a)) match well with JCPDS data of $LaFeO_3$ [JCPDS PDF# 37- 1493]. They also reveal that all the synthesized multiferroics are monophasic exhibiting orthogonal symmetry with space group *Pnma* (Space group No: 62) [Aroyo, 2016]. In the XRD patterns of $La_{1-x}Sr_xFeO_3$ multiferroics, it is expected that the shift of XRD peaks towards the lower angle side because of the ionic radius of Sr^{2+} (1.18 Å) [Shannon, 1976] is much larger than that of La^{3+} (1.03 Å) [Shannon, 1976]. But the XRD peaks shift towards the higher angle side with respect to the XRD peaks of x=0.05 (Figure 2.7 (a)). A similar trend of XRD peaks shift in Sr-substituted $LaFeO_3$ has been observed by the other researchers also [Richens, 1997, Nguyen et al., 2014]. The enlarged XRD peak corresponding to (121) crystallographic plane (Figure 2.7(b)) depicts the right shift of XRD peaks for all the Sr concentrations. The reason for this right shift can be explained as follows: The replacement of La^{3+} ions by Sr^{2+} ions induces the charge compensation of Fe^{3+} ions in the crystal lattice [Richens, 1997, Nguyen et al., 2014]. Consequently, Fe^{3+} ions with a larger radius of 0.65 Å were oxidized to Fe^{4+} ions with a smaller radiusof 0.58 Å [Richens, 1997], which results in the right shift of XRD peaks. Thus, XRD results indicate that the substitution of Sr^{2+} ions at the lattice site of La^{3+} ions decreases the lattice spacing, thereby leads to the lattice shrinkage.

Table 3.2 *Comparison of refined orthorhombic unit cell parameters for the synthesized multiferroics*

Samples	Conc. (x)	Lattice parameters a (Å)	b (Å)	c (Å)	Unit cell volume (Å³)
LCFO	0.00	5.5633(27)	7.8546(23)	5.5564(9)	242.80(7)
	0.03	5.5619(19)	7.8543(5)	5.5513(6)	242.51(5)
	0.06	5.5617(28)	7.8521(23)	5.5512(10)	242.43(9)
	0.09	5.5616(35)	7.8485(29)	5.5528(12)	242.38(1)
	0.12	5.5612(30)	7.8551(25)	5.5510(10)	242.49(7)
LZFO	0.00	5.5629(2)	7.8551(1)	5.5531(1)	242.66(4)
	0.05	5.5683(9)	7.8486(7)	5.5481(6)	242.47(3)
	0.15	5.5645(4)	7.8553(3)	5.5507(3)	242.63(9)
	0.25	5.5655(3)	7.8531(2)	5.5531(2)	242.71(7)
LAFO	0.05	5.5643(4)	7.8553(11)	5.5611(6)	243.07(2)
	0.15	5.5567(3)	7.8493(3)	5.5577(2)	242.41(17)
	0.25	5.5432(2)	7.8401(2)	5.5437(5)	240.93(2)
LSFO	0.05	5.5551(4)	7.8452(4)	5.5532(3)	242.02(3)
	0.10	5.5516(4)	7.8443(6)	5.5507(2)	241.73(6)
	0.15	5.5421(3)	7.8410(5)	5.5435(5)	240.90(2)
	0.20	5.5313(4)	7.8317(7)	5.5389(3)	239.95(7)

LCFO - $La_{1-x}Ce_xFeO_3$ LZFO - $La_{1-x}Zn_xFeO_3$ LAFO - $La_{1-x}Al_xFeO_3$ LSFO - $La_{1-x}Sr_xFeO_3$

Conc. – concentration

The structural refinement of $La_{1-x}Sr_xFeO_3$ was carried out with orthorhombic structure having space group *Pnma* (Space group No: 62) [Aroyo, 2016] with 4 moleculesper unit cell. In the orthorhombic system, the atomic positional coordinates were taken as (0.0295, 0.25, 0.9951) for La/Sr, (0.5, 0, 0) for Fe, (0.488, 0.25, 0.0707) for apex oxygen O1 and (0.2831, 0.0387, 0.7168) for planar oxygen O2 [Wyckoff, 1963]. During Rietveld refinement [Rietveld, 1969], the lattice constants, scale factor, thermal parameters, pseudo-Voigt, asymmetry, peak shift, background profile shape and preferred orientation were refined to minimize the difference between the calculated profile with the observed one. Figures 2.8 (a)-(d) depict the fitted powder XRD profiles of $La_{1-x}Sr_xFeO_3$ (x=0.05, 0.10, 0.15 & 0.20) multiferroics. The structural and profile fitted parameters obtained from the Rietveld refinement method [Rietveld, 1969] (Table 2.4) show that the unit cell parameters and cell volume decrease with the increase in Sr concentration. This is attributed to the fact that the substitution of larger ionic radius Sr^{2+} ions in the place of smaller ionic radius La^{3+} ions induces the charge compensation of Fe^{3+} ions in the crystal lattice. To compensate the charges in the crystal lattice of $La_{1-x}Sr_xFeO_3$ multiferroics, Fe^{3+} ions with a radius of 0.65 Å [Shannon, 1976] were oxidized to Fe^{4+} ions with a smaller radius of 0.58 Å [Shannon, 1976], which results in lattice shrinkage [Richens, 1997, Nguyen et al., 2014]. The perfect profile fitting of all the synthesized multiferroics confirms that the measured data (crosses) and calculated diffraction profiles (continuous solid line) are in good agreement with each other. Hence, the structure factors evolved from Rietveld refinement [Rietveld, 1969] are considered as the most precise, which have been used for the charge density analysis.

3.4   Grain size analysis of lanthanum orthoferrite-type multiferroics

The average grain size of all the synthesized lanthanum orthoferrite-typemultiferroics has been determined using the Scherrer formula [Cullity, 2001, Gorelov et al., 2011],

$$t = \frac{0.9\lambda}{\beta \cos\theta}$$

where, t is the average grain size (average size of coherently diffracting domains), λ is the wavelength of X-ray used (1.54056 Å), β is the full width at half maximum (FWHM) in radians for the prominent intensity peak and θ is the corresponding Bragg angle, through the GRAIN software program [Saravanan, 2008].

The average grain size of $La_{1-x}Ce_xFeO_3$ (x=0.00, 0.03, 0.06, 0.09 & 0.12) multiferroics are 61.8(4.5) nm, 79.7(3.1) nm, 59.1(4.6) nm, 57.9(2.4) nm and 88.1(1.3) nm for x=0.00, 0.03, 0.06, 0.09 & 0.12 respectively.

The average grain size of $La_{1-x}Zn_xFeO_3$ (x=0.00, 0.05, 0.15 & 0.25) multiferroicsare 36.7(1) nm, 26.9(0.3) nm, 27.8(0.9) nm and 25.9(0.5) nm for x=0.00, 0.05, 0.15 & 0.25 respectively.

The average grain size of $La_{1-x}Al_xFeO_3$ (x=0.05, 0.15 & 0.25) multiferroics are37.7(1.7) nm, 36.1(1.4) nm and 26.1(0.6) nm for x=0.05, 0.15 & 0.25 respectively.

The average grain size of $La_{1-x}Sr_xFeO_3$ (x=0.05, 0.10, 0.15 & 0.20) multiferroicsare 26.1(0.8) nm, 24.8(1.7) nm, 22.2(3.5) nm and 21.4(1.2) nm for x=0.05, 0.10, 0.15 & 0.20 respectively.

The average grain size range of all the synthesized lanthanum orthoferrite-type multiferroics determined using the Scherrer formula [Cullity, 2001, Gorelov et al., 2011] have been given in Table 3.3.

Table 3.3 *Average grain size range of lanthanum orthoferrite-type multiferroics from XRD*

Samples	Average grain size range (nm)
LCFO	57.9(2.4) - 88.1(1.3)
LZFO	25.9(0.5) - 36.7(1)
LAFO	26.1(0.6) - 37.7(1.7)
LSFO	21.4(1.2) - 26.1(0.8)

LCFO - $La_{1-x}Ce_xFeO_3$
LZFO - $La_{1-x}Zn_xFeO_3$
LAFO - $La_{1-x}Al_xFeO_3$
LSFO - $La_{1-x}Sr_xFeO_3$

3.5 Microstructure and elemental analysis of lanthanum orthoferrite-type multiferroics

The surface morphology and microstructure of all the synthesized multiferroics have been analyzed by SEM images recorded with various magnifications ranging from ×500 to ×15000. The average particle size has also been determined from SEM micrographs. The elemental compositions of all the synthesized multiferroics have been investigated qualitatively and quantitatively from energy dispersive spectroscopy (EDS) measurements.

3.5.1 Ce-substituted lanthanum orthoferrites - $La_{1-x}Ce_xFeO_3$

The SEM images of the prepared ($La_{1-x}Ce_xFeO_3$) multiferroics (Figures 2.9 (a)-(e)) recorded with magnification x7000 show that the formed crystals have numerous deagglomerated particles that are heterogeneously distributed with irregular shapes and different sizes. The average particle size for the pure and Ce-substituted samples is in the range from 0.86(0.09) μm to 1.09(0.1) μm.

The EDS spectra of $La_{1-x}Ce_xFeO_3$ (Figures 2.10 (a)-(e)) depict various peaks belonging to the constituent elements of the samples. The numerical values of atomic and weight percentages of the various elements present in the multiferroics (Table 2.5) clearly reveal that the prepared $La_{1-x}Ce_xFeO_3$ contains atomic species namely La, Ce, Fe, and O without any impurities.

3.5.2   Zn-substituted lanthanum orthoferrites - $La_{1-x}Zn_xFeO_3$

The SEM micrographs of $La_{1-x}Zn_xFeO_3$ (Figures 2.11(a)-(d)) recorded with magnification x7000 show particles have been deagglomerated and heterogeneously distributed in all the samples with irregular shapes and different sizes. The average particle size for the prepared $La_{1-x}Zn_xFeO_3$ with Zn concentrations x=0.00, 0.05, 0.15 and 0.25 is found to be in the range from 0.91(0.4) μm to 1.34(0.6) μm. The average particle size decreases non-monotonically with the increase in the Zn concentration, which is almost consistent with the trend of variation in grain size obtained from XRD results.

The EDS spectra of $La_{1-x}Zn_xFeO_3$ (Figures 2.12(a)-(d)) show the various peaks corresponding to the atomic species such as La, Zn, Fe, and O, which confirm the presence of constituent elements in all the samples. The numerical values of atomic and weight percentages of the constituent elements in $La_{1-x}Zn_xFeO_3$ (Table 2.6) reflect that the La concentration decreases monotonically with increasing Zn concentration as expected from XRD results. The numerical values in Table 2.6 also confirm that no other impurities are present in the samples.

3.5.3   Al-substituted lanthanum orthoferrites - $La_{1-x}Al_xFeO_3$

The SEM micrographs of $La_{1-x}Al_xFeO_3$ (Figures 2.13(a)-(c)) recorded with magnification x10000 show the deagglomerated particles having different shapes with irregular boundaries and size in the micrometer range. The average particle size estimated from the SEM photographs of the samples is found to be in the range from 0.64(0.2) μm to 0.82(0.18) μm.

The EDS spectra of $La_{1-x}Al_xFeO_3$ (Figures 2.14(a)-(c)) depict the characteristic X-ray peaks of La, Al, Fe, and O elements, which are the constituents in the synthesized $La_{1-x}Al_xFeO_3$ multiferroics. The EDS spectra (Figures 2.14(a)-(c)) also depict a small intense peak at 1.75 KeV corresponding to the Si-Kα characteristic line, which is due to the silicon substrate on the mounting base of the Si-Li detector [Ibers and Hamilton, 1974]. The numerical values of atomic and weight percentages of the constituent elements in $La_{1-x}Al_xFeO_3$ multiferroics (Table 2.7) reflect that the La concentration decreases systematically with increasing Al concentration as expected, and they confirm the purity of the samples.

3.5.4   Sr-substituted lanthanum orthoferrites - $La_{1-x}Sr_xFeO_3$

The SEM images of the $La_{1-x}Sr_xFeO_3$ multiferroics corresponding to the magnification x10000 (Figures 2.15(a)-(d)) depict that the formed powder crystals have particles that are heterogeneously distributed with irregular shapes and sizes. The average particle size for

$La_{1-x}Sr_xFeO_3$ ranges from 0.68(0.05) μm to 0.88(0.1) μm. The particle size decreases with the increase in the substitution of Sr^{2+} ions. The decreasing trend of particle size with increasing substitution of Sr is consistent with the grain size obtained from the results of XRD. The decrease in the average particle/grain size of the prepared $La_{1-x}Sr_xFeO_3$ multiferroics is due to the occurrence of charge compensation of Fe^{3+} ions inthe crystal lattice, wherein the Fe^{3+} ions were oxidized to Fe^{4+} ions resulting decrease in the number of Fe^{3+} ions.

Table 3.4 *The average particle size of lanthanum orthoferrite-type multiferroics*

Samples	Conc.(x)	Average particle size (μm)
LCFO	0.00	1.09(0.1)
	0.03	0.97(0.08)
	0.06	0.95(0.1)
	0.09	0.86(0.09)
	0.12	1.06(0.06)
LZFO	0.00	1.34(0.6)
	0.05	0.91(0.4)
	0.15	1.29(0.23)
	0.25	1.30(0.49)
LAFO	0.05	0.82(0.18)
	0.15	0.71(0.12)
	0.25	0.64(0.2)
LSFO	0.05	0.88(0.1)
	0.10	0.80(0.06)
	0.15	0.69(0.3)
	0.20	0.68(0.05)

LCFO - $La_{1-x}Ce_xFeO_3$
LZFO - $La_{1-x}Zn_xFeO_3$
LAFO - $La_{1-x}Al_xFeO_3$
LSFO - $La_{1-x}Sr_xFeO_3$
Conc. - concentration

The EDS spectra of $La_{1-x}Sr_xFeO_3$ multiferroics (Figures 2.16(a)-(d)) clearly depict the peaks belonging to the constituent ions such as La, Sr, Fe, and O of the synthesized multiferroics. No other impurities were detected in the energy dispersive X-ray spectrum. The numerical values of atomic and weight percentages of various elements (Table 2.8) clearly reveal that the prepared samples contain constituent atomic species with no other impurities. The numerical values in Table 2.8 also reflect that La:Sr:Fe ratio follows the expected trend except for the sample with x=0.1. This may be due to inhomogeneity in the grain structure caused by the improper mixture of the sample. As a result, the elemental composition percentage may vary from point to point. In this work, the EDS measurement has been performed at only one point of the sample. The elemental composition at that particular point of the sample may not be in proper proportion and that may be reflected in the EDS measurement. However, the increasing trend of Sr in Table 2.8 confirms the incorporation of the element Sr in the host matrix.

The average particle size of the synthesized lanthanum orthoferrite-type multiferroics estimated from the SEM images have been summarized in Table 3.4.

3.6 Charge density distribution analysis of lanthanum orthoferrite-type multiferroics

The precise electronic structure, inter-atomic chemical bonding and charge density distribution in the unit cell of lanthanum orthoferrite-type multiferroics have been effectively analyzed using the Maximum Entropy Method (MEM) [Collins, 1982]. The numerical MEM [Collins, 1982] computations have been carried out by partitioning the orthorhombic unit cell into $48 \times 72 \times 48$ pixels along the a, b and c axes by employing the software program PRIMA [Izumi, 2002]. The resultant 3D and 2D charge density distribution are plotted with the help of the visualization software package VESTA [Momma, 2008].

3.6.1 Ce-substituted lanthanum orthoferrites - $La_{1-x}Ce_xFeO_3$

The three-dimensional (3D) charge density maps (Figures 2.17 (a)-(e)), which have been constructed with the similar iso-surface level of 3 e/$Å^3$ show the charge density distribution between the atoms in the unit cell of $La_{1-x}Ce_xFeO_3$ (x=0.00, 0.03, 0.06, 0.09 & 0.12) multiferroics. The crystallographic positions of La, Fe and O atoms are clearly seen with electron clouds. The 3D view of $La_{1-x}Ce_xFeO_3$ (x=0.00) unit cell with FeO_6 octahedra (Figure 2.18) illustrates that in the orthorhombic structure, the Fe^{3+} ion lies at the center of octahedron formed by six O^{2-} ions, while La^{3+} ions are at the space between the FeO_6 octahedron.

The structural stability due to the twist of FeO_6 octahedra, which is caused by the distortion of structure to accommodate the various cations in the host lattice of the $LaFeO_3$ system has been estimated using the Goldschmidt tolerance factor (t) relation [Goldschmidt, 1926]. The tolerance factor (t) for $La_{1-x}Ce_xFeO_3$ multiferroics is 0.841, 0.839, 0.838, 0.836 and 0.834 for x=0.00, 0.03, 0.06, 0.09 and 0.12 respectively. The tolerance factor (t) decreases for the increase in the substitution of Ce. This indicates the twisting of the FeO_6 octahedral framework due to the distortion of structure to accommodate the smaller cation

Ce^{4+} (As the ionic radius of La^{3+} (r_{La} =1.03 Å) [Shannon, 1976] is larger than that of Ce^{4+} (r_{Ce}=0.87 Å)) [Shannon, 1976].

The twist of $< FeO_6 >$ octahedral framework due to the substitution of Ce in $LaFeO_3$ has been analyzed quantitatively using MEM [Collins, 1982] based charge density studies. Table 2.9 summarizes the variation in bond angles (out-plane bond angle (Fe–O1–Fe) and in-plane bond angle (Fe–O2–Fe)) and bond lengths (apical atoms distance (Fe–O1) and planar atoms distance (Fe–O2)) in the Fe-O sub-lattice of $La_{1-x}Ce_xFeO_3$, due to FeO_6 octahedral twist which is caused to accommodate the smaller cation Ce^{4+} in the host lattice.

The three-dimensional unit cell of $La_{1-x}Ce_xFeO_3$ (x=0.00) with (100) plane shaded (Figure 2.19(a)) and the two-dimensional charge density distribution of $La_{1-x}Ce_xFeO_3$ (x=0.00) on (100) plane (Figure 2.19(b)) with the enlarged view of the Fe-O1 bond (Figures 2.19 (c)-(g)) illustrates the interaction between the Fe and O1 atoms. To draw the 2D charge density maps for the Fe-O1 bond (Figures 2.19 (b)-(g)) the contour lines on the

(100) Miller plane are drawn in the range 0-1.2 $e/Å^3$ with 0.08 $e/Å^3$ interval. The enlarged view of the Fe-O1 bond for various Ce concentrations x=0.00, 0.03, 0.06, 0.09 & 0.12 (Figures 2.19 (c)-(g)), show that the contour lines around the Fe atom are elongated towards the oxygen atom in the bonding region, which indicates that there is an accumulation of charges towards the O1 site. This observation confirms that the bonding between the Fe and O1 atoms is predominantly covalent. Figures 2.19 (c)-(g), also show that for x=0.06, the accumulation of charges towards the O1 site is more when compared to other Ce-substituted samples, which indicates that the Fe-O1 bond is more covalent in this substitution level.

The three-dimensional unit cell of $La_{1-x}Ce_xFeO_3$ (x=0.00) with (200) plane shaded (Figure 2.20(a)) and the two dimensional charge density distribution of $La_{1-x}Ce_xFeO_3$ (x=0.00) on (200) plane (Figure 2.20 (b)) with the enlarged view of the La-O1 bond (Figures 2.20 (c)-(g)) illustrates the interaction between the La and O1 atoms. To draw the 2D charge density maps for the La-O1 bond (Figures 2.20 (b)-(g)) the contour lines on the (200) Miller plane are drawn in the range 0-1.3 $e/Å^3$ with 0.09 $e/Å^3$ interval. The enlarged view of the La-O1 bond for various Ce concentrations x=0.00, 0.03, 0.06, 0.09 & 0.12 (Figures 2.20 (c)-(g)) shows that the contour lines around the La atom are elongated towards the oxygen atom and also overlapped in the bonding region, which indicates the localization or accumulation of charges towards the O1 site. The accumulation of charges, which arises due to the spatial distribution of charges from neighboring atoms, indicates that the bond La-O1 is equally covalent. Figures 2.20 (c)-(g), also show that for x=0.06, the accumulation of charges towards the O1 site is more when compared to other Ce-substituted samples, which indicates that the bond La-O1 is more covalent in this substitution level.

To analyze the bonding features quantitatively, the one dimensional charge density profiles along the bonds Fe-O1 and La-O1 are drawn as shown in Figure (2.21) and Figure (2.22) respectively. The bond lengths and bond critical point (BCP) charge densities along the F-O1 and La-O1 bonds of the synthesized LCFO-type multiferroics are tabulated in Table 2.10. From Table 2.10, it is evident that for the pristine sample, the bond critical point charge density along the Fe-O1 bond is found to be high as 0.9251 e/Å^3. The high value of charge density confirms that the bond Fe-O1 is predominantly covalent. Again from Table 2.10, it is seen that for the pristine sample, the bond critical point charge density along the La-O1 bond is also found to be high as 0.9811 e/Å^3 due to spatial distribution of charges. The high value of charge density confirms that the bond La-O1 is equally covalent in nature with slightly more charge density. Among the Ce-substituted samples, for x=0.06, the bond critical point charge densities along the Fe-O1 bond and La-O1 bond are found to be relatively high as 0.834 e/Å^3 and 1.0741 e/Å^3, which confirms that the bond Fe-O1 is more covalent and the bond La-O1 is also more covalent with slightly more charge density in this substitution level.

3.6.2 Zn-substituted lanthanum orthoferrites - La$_{1-x}$Zn$_x$FeO$_3$

The 3D charge density distribution maps of La$_{1-x}$Zn$_x$FeO$_3$ (Figures 2.23 (a)-(d)) have been constructed with the similar iso-surface level of 3 e/Å^3, and they reflect the orthorhombic structure for all the samples. From the 3D charge density maps, the crystallographic sites of the La, Fe and O atoms and electron clouds around them are seen clearly. Figure 2.24 illustrates the 3D view of the synthesized La$_{1-x}$Zn$_x$FeO$_3$ (x=0.00) unit cell with FeO$_6$ octahedra, where the Fe^{3+} ion lies at the center of octahedron formed by six O^{2-} ions, while La^{3+} ion occupies the space between the FeO$_6$ octahedra. The Goldschmidt tolerance factor (t) [Goldschmidt, 1926] for La$_{1-x}$Zn$_x$FeO$_3$multiferroics is found to be 0.8409, 0.8358, 0.8257 and 0.8156 for x=0.00, 0.05, 0.15 and 0.25 respectively. The tolerance factor (t) of La$_{1-x}$Zn$_x$FeO$_3$ multiferroics decreases for theincreasing incorporation of Zn^{2+} in the La^{3+} site of LaFeO$_3$. The deviation of tolerance factor signifies the tilting of the FeO$_6$ octahedral framework due to the distortion of orthorhombic structure to accommodate the smaller cation Zn^{2+} (As the ionic radius of La^{3+} (r$_{La}$=1.03 Å) is larger than that of Zn^{2+} (r$_{Zn}$=0.74 Å)) [Shannon, 1976].

The degree of distortion of FeO$_6$ octahedra due to the replacement of La^{3+} by Zn^{2+}ion has been examined for the synthesized La$_{1-x}$Zn$_x$FeO$_3$ using MEM. Table 2.11 showsthe variation in bond angles ((Fe–O1–Fe) and (Fe–O2–Fe)) and bond lengths ((Fe–O2)and (Fe–O1)) in the Fe-O sub-lattice of La$_{1-x}$Zn$_x$FeO$_3$ caused by the FeO$_6$ octahedral twist.The three-dimensional unit cell of La$_{1-x}$Zn$_x$FeO$_3$ (x=0.00) with (100) plane shaded (Figure 2.25(a)) and the two dimensional charge density distribution of La$_{1-x}$Zn$_x$FeO$_3$ (x=0.00) on (100) plane (Figure 2.25(b)) with the enlarged view of the Fe-O1 bond (Figures 2.25 (c)-(f)) show the interaction between the Fe and O1 atoms. The 2D charge density contour maps on the (100) plane for the Fe-O1 bond (Figures 2.25 (b)-(f)) are drawn in the range 0 – 1 e/$Å^3$ with an interval of 0.11 e/$Å^3$. From the enlarged view of the bonding between the Fe and O1 atoms for x=0.00, 0.05, 0.15 and 0.25 (Figures 2.25 (c)-(f)), it is seen that the contour lines around the Fe atom are elongated towards the O1 atom, which indicates the sharing of charges between the Fe and O1 atoms. This kind of distribution of contour lines in the bonding region reveals that a predominant covalent nature exists between the Fe and O1 atoms. Figures 2.25 (c)-(f) also illustrate that for Zn-substituted samples the accumulation of charges towards the O1 site is less than that for x=0.00. Among the Zn-substituted samples, for x=0.25, the accumulation of charges towards the O1 site is found to be more when compared to others, which indicates that thebond Fe-O1 is more covalent in this substitution level.

The three-dimensional unit cell of $La_{1-x}Zn_xFeO_3$ (x=0.00) with (200) plane shaded (Figure 2.26(a)) and the two dimensional charge density distribution of $La_{1-x}Zn_xFeO_3$ (x=0.00) on (200) plane (Figure 2.26(b)) with the enlarged view of the La-O1 bond (Figures 2.26 (c)-(f)) show the interaction between La and O1 atoms. The 2D charge density contour maps on the (200) plane for the La-O1 bond (Figures 2.26 (b)-(f)) are drawn in the range 0 – 1.4 e/$Å^3$ with an interval of 0.09 e/$Å^3$. From the enlarged view of the bonding between the La and O1 atoms for x=0.00, 0.05, 0.15, and 0.25 (Figures 2.26 (c)-(f)), it is observed that the contour lines around the La atom are crowded and overlapped in the bonding region, which indicates the localization or accumulation of charges towards the O1 site. The accumulation of charges, which arises due to the spatial distribution of charges from neighboring atoms, indicates that the bond La-O1 is equally covalent. Figures 2.26 (c)-(f) also reveal that among the prepared samples, the accumulation of charges for Zn-substituted samples is less than that for x=0.00. Among the Zn-substituted samples, for x=0.25, the accumulation of charges is found to be more when compared to others, which indicates that the bond La-O1 is more covalent in this substitution level.

The quantitative analysis of the above-said primary chemical bonding has beendone by drawing the 1D charge density line profiles (Figure (2.27) and Figure (2.28)) and hence by estimating the charge density values at the bond critical point (BCP) along the primary bonds Fe-O1 and La-O1 of $La_{1-x}Zn_xFeO_3$. The bond lengths and bond critical point (BCP) charge densities along the Fe-O1 and La-O1 bonds of the synthesized $La_{1-x}Zn_xFeO_3$ are tabulated in Table 2.12. From Table 2.12, it is seen that for the pure sample, the charge density is found to be high as 0.6712 e/$Å^3$ at the bond critical point of the bond Fe–O1. The high value of charge density confirms that the bond Fe-O1 is predominantly covalent. Again, from Table 2.12, it is observed that for the La-O1 bond the charge density value is found to be high as 0.6872 e/$Å^3$ due to the spatial distribution of charges. The high value of charge density confirms that the bond La-O1 is equally covalent with slightly more charge density. For the Zn-substituted samples, the BCP charge density values along the primary bonds Fe-O1 and La-O1 are found to be less than those for x=0.00. Among the Zn-substituted samples, for the sample with Zn concentration x=0.25, the BCP charge density values along the primary bonds Fe-O1 and La-O1 are found to be relatively high as 0.5929 e/$Å^3$ and 0.5671 e/$Å^3$ respectively, which confirms that the bond Fe-O1 is more covalent and the bond La-O1 is also more covalent with slightly more charge density in this substitution level.

3.6.3 Al-substituted lanthanum orthoferrites - $La_{1-x}Al_xFeO_3$

The three-dimensional (3D) charge density iso-surfaces in the unit cell of $La_{1-x}Al_xFeO_3$ (x=0.05, 0.15 & 0.25) multiferroics (Figures 2.29 (a)-(c)) have been constructed with the similar iso-surface level of 3 e/$Å^3$. The 3D visualizations of charge density iso-surfaces of $La_{1-x}Al_xFeO_3$ (Figures 2.29 (a)-(c)) clearly depict the lattice sites of La, Fe, and O atoms and charge clouds in their valence region. Also, these 3D charge density maps confirm the orthorhombic structure of the samples. Figure 2.30 illustrates the 3D network of FeO_6 octahedron in the orthorhombic unit cell of the Al-substituted sample ($La_{1-x}Al_xFeO_3$ (x=0.05)).

The tolerance factor for $La_{1-x}Al_xFeO_3$ multiferroics is found to be 0.8323, 0.8151,0.7979 and 0.7807 for Al concentration 0.05, 0.15, 0.25 and 0.35 respectively and itdecreases with the increasing substitution of Al. The decreasing tolerance factorrepresents the twisting of < FeO_6 > octahedral framework to accommodate the smallercation Al^{3+} (r_{Al}=0.535 Å) in the place of larger cation La^{3+} (r_{La}=1.03 Å) [Shannon, 1976].

The twist of < FeO_6 > octahedra has been analyzed quantitatively by extracting

the bond angles and bond lengths in the Fe-O sub-lattice of $La_{1-x}Al_xFeO_3$ using MEM-based charge density studies. Table 2.13 gives the variation in (Fe–O1–Fe), (Fe–O2–Fe) bond angles and (Fe–O1), (Fe–O2) bond lengths in the Fe-O sub-lattice of $La_{1-x}Al_xFeO_3$, due to the FeO_6 octahedral twist caused by the substitution of Al in $LaFeO_3$.

To examine the nature of the primary bonds Fe-O1 and La-O1 qualitatively, the 2D charge density distribution maps are drawn on two different Miller planes (100) and (200). The three-dimensional unit cell of $La_{1-x}Al_xFeO_3$ (x=0.05) with (100) plane shaded (Figure 2.31(a)) and the two-dimensional charge density distribution of $La_{1-x}Al_xFeO_3$ (x=0.05) on the (100) crystallographic plane (Figure 2.31(b)) with the enlarged view of the Fe-O1 bond (Figures 2.31 (c)-(e)) illustrate the interaction between the Fe and O1 atoms. The 2D charge density maps on the (100) Miller plane for the Fe-O1 bond (Figures 2.31 (b)-(e)) are drawn in the contour range of $0 - 1.7$ e/Å^3 with the contour interval of 0.12 e/Å^3. The enlarged view of the Fe-O1 bond for various Al concentrations x=0.05, 0.15 & 0.25 (Figures 2.31 (c)-(e)), shows that the contour lines around the Fe atom are elongated towards the oxygen atom in the bonding region, which indicates that the localization or sharing of charges between the Fe and O1 atoms. The sharing of charges reveals a predominant covalent nature exists between the Fe and O1 atoms. Figures 2.31 (c)-(e) also reveal that for x=0.25, the sharing of charges is more when compared to other Al-substituted samples, which indicates that the bond Fe-O1 is more covalent in this substitution level.

The three-dimensional unit cell of $La_{1-x}Al_xFeO_3$ (x=0.05) with (200) plane shaded (Figure 2.32(a)) and the two dimensional charge density distribution of $La_{1-x}Al_xFeO_3$ (x=0.05) on the (200) crystallographic plane (Figure 2.32(b)) with the enlarged view of the La-O1 bond (Figures 2.32 (c)-(e)) illustrate the interaction between La and O1 atoms. The 2D charge density maps on the (200) Miller plane for the La-O1 bond (Figures 2.32 (b)-(e)) are drawn in the contour range of $0-1.2$ e/Å^3 with the contour interval of 0.09 e/Å^3. Figures 2.32 (c)-(e) shows the bounded contour lines around the La atom, which indicates that there is delocalization or no sharing of charges between the La and O1 atoms. The delocalization of charges in the bonding region reveals that a less covalent nature exists between La and O1 atoms. Figures 2.32 (c)-(e) also show that for the Al concentration x=0.15, the contour lines are slightly elongated from La atom to O1 atom, which reflects the localization or accumulation of charges towards the O1 site. This observation indicates that the bond La-O1 is more covalent in this substitution level.

To understand the strength and nature of the above-said chemical bonds, the one-dimensional (1D) charge density profiles (Figures (2.33) and (2.34)) along the Fe-O1 bond and La-O1 bond are drawn and the charge density values at the bond critical point (BCP) along the primary bonds Fe-O1 and La-O1 of $La_{1-x}Al_xFeO_3$ are estimated. The bond lengths and estimated BCP charge densities along the bonds Fe-O1 and La-O1 are given in Table 2.14. From Table 2.14, it is evident that for the Fe-O1 bond, the bond critical point charge density values vary from 0.876 e/Å^3 to 1.015 e/Å^3. The high values of BCP charge density confirm that the bond Fe-O1 is predominantly covalent. Among the prepared samples, for x=0.25, the BCP charge density value is found to be relatively high as 1.015 e/Å^3, which confirms that the bond Fe-O1 is more covalent in this substitution level. Again, from Table 2.14, it is seen that for the La-O1 bond, the BCP charge densities vary from 0.3175 e/Å^3 to 0.4699 e/Å^3. The low values of BCP charge density declare that the bond La-O1 is less covalent with slightly less charge density. Among the prepared samples, for x=0.15, the BCP charge density value is relatively high as 0.4699 e/Å^3, which confirms that the bond La-O1 is more covalent with slightly more charge density in this substitution level.

3.6.4   Sr-substituted lanthanum orthoferrites - $La_{1-x}Sr_xFeO_3$

The three-dimensional (3D) MEM pictures of charge density distribution in the unit cell of $La_{1-x}Sr_xFeO_3$ multiferroics (Figures 2.35 (a)-(d)) have been drawn with the similar iso-surface level of 2 $e/Å^3$. From the three-dimensional charge density maps, the atomic sites of the La, Fe and O atoms with electron clouds around them are clearly visualized and these maps confirm the orthorhombic structure of the prepared $La_{1-x}Sr_xFeO_3$ multiferroics. Figure 2.36 illustrates the 3D network of FeO_6 octahedron in the orthorhombic unit cell of Sr-substituted sample ($La_{1-x}Sr_xFeO_3$ (x=0.05)).

The Goldschmidt tolerance factor (t) [Goldschmidt, 1926] for the $La_{1-x}Sr_xFeO_3$ multiferroics is 0.8434, 0.8460, 0.8486 and 0.8511 for Sr concentration 0.05, 0.1, 0.15 and 0.2 respectively. The tolerance factor (t) increases for the increasing substitution of Sr, which signifies the twisting of the FeO_6 octahedral framework due to the distortion of structure to accommodate the larger cation Sr^{2+} (As the ionic radius of La^{3+} (r_{La}=1.03 Å) is smaller than Sr^{2+} (r_{Sr}=1.18 Å)) [Shannon, 1976] in the host lattice.

The twist of the FeO_6 octahedral framework due to Sr substitution has been investigated through MEM studies. Table 2.15 presents the change in bond angles ((Fe-O1-Fe) and (Fe-O2-Fe)) and bond lengths ((Fe-O1) and (Fe-O2)) in the Fe-O sub-lattice caused by the twist of FeO_6 octahedron of $La_{1-x}Sr_xFeO_3$ to accommodate Sr^{2+} in the host lattice.

The 3D unit cell of $La_{1-x}Sr_xFeO_3$ (x=0.05) with (100) plane shaded (Figure 2.37(a)) and the two-dimensional charge density distribution of $La_{1-x}Sr_xFeO_3$ (x=0.05) on the (100) crystallographic plane (Figure 2.37(b)) with the enlarged view of the Fe-O1 bond (Figures 2.37 (c)-(f)) illustrates the interaction between the Fe and O1 atoms. The 2D charge density maps on the (100) Miller plane for the Fe-O1 bond (Figures 2.37 (b)-(f)) are drawn in the contour range of 0 - 1.2 $e/Å^3$ with 0.09 $e/Å^3$ interval. The enlarged view of Fe-O1 bond for various Sr concentrations x=0.05, 0.10, 0.15 & 0.25 (Figures 2.37 (c)-(f)), shows the elongation of contour lines from the Fe site to the O1 site in the bonding region, which indicates that there is localization or sharing of charges between the Fe and O1 atoms. The sharing of charges reveals that a predominant covalent nature exists between the Fe and O1 atoms. Figures 2.37 (c)-(f) also depict that for the sample with Sr concentration x=0.05, the accumulation of charges towards the O1 site is more when compared to other Sr-substituted samples, which indicates that the bondFe-O1 is more covalent in this substitution level.

The 3D unit cell of $La_{1-x}Sr_xFeO_3$ (x=0.05) with (200) plane shaded (Figure 2.38(a)) and the two-dimensional charge density distribution of $La_{1-x}Sr_xFeO_3$ (x=0.05) on the (200) crystallographic plane (Figure 2.38(b) with the enlarged view of the La-O1 bond (Figures 2.38 (c)-(f)) illustrate the interaction between the La and O1 atoms. The 2D charge density maps on the (200) Miller plane for the La-O1 bond (Figures 2.38 (b)-(f)) are drawn in the contour range of 0 - 1.2 $e/Å^3$ with 0.09 $e/Å^3$ interval. The enlarged view of La-O1 bond for various Sr concentrations x=0.05, 0.10, 0.15 & 0.25 (Figures 2.38 (c)-(f)) depicts that the contour lines around the La atom are crowded and overlapped in the bonding region. The overlapped contour lines indicate the localization or accumulation of charges towards the O1 site which arises due to the spatial distribution of charges from neighboring atoms. The localization of charges from the La site to the O1 site reveals that the bond La-O1 is equally covalent. Figures 2.38 (c)-(f) also reveal that for x=0.05, the accumulation of charges towards the O1 site is more when compared to other Sr-substituted samples, which indicates that the bond La-O1 is more covalent in this substitution level.

Precise electron density variation along the primary bonds Fe-O1 and La-O1 due to Sr substitution has been analyzed by drawing the one-dimensional charge density profiles (Figures 2.39 and 2.40) along the bonds Fe-O1 and La-O1 and by extracting the numerical values of charge density at the bond critical point. The bond lengths and bond critical point charge densities of Fe-O1 and La-O1 bonds of the synthesized $La_{1-x}Sr_xFeO_3$ multiferroics are given in Table 2.16. The estimated charge densities at the bond critical point along the Fe-O1 bond of the synthesized samples are found to be high, and they are found to be in the range of 0.7782 $e/Å^3$ to 0.8693 $e/Å^3$. The high value of charge density declares that the bond Fe-O1 is predominantly covalent in nature. Again, from Table 2.16, it is seen that for the La-O1 bond, due to the spatial distribution of charges from neighboring atoms, the charge density values at the bond critical point along the La-O1 bond of all the samples are high, and they are found to be in the range of 0.5281 $e/Å^3$ to 1.0203 $e/Å^3$, which declares that the bond La-O1 is equally covalent. Among the prepared samples, for the sample with x=0.05, the values of charge density along the bonds Fe-O1 and La-O1 are found to be relatively high as 0.8693 $e/Å^3$ and 1.0203 $e/Å^3$, which confirms that the bond Fe-O1 is more covalent and the bond La-O1 is also more covalent with slightly more charge density in this substitution level.

For the synthesized lanthanum orthoferrite-type multiferroics the tolerance factor t is < 0.96, which indicates that all the samples distort into an orthorhombic structure. The change in bond angles and bond lengths in the Fe-O sub-lattice of lanthanum orthoferrite- type multiferroics due to < FeO_6 > octahedral twist, which is caused by the substitution of Ce^{4+}, Zn^{2+}, Al^{3+} and Sr^{2+} cations are summarized in Table 3.5. The variation in bond angles and bond lengths describes the distortion in FeO_6 octahedra. The bond critical point charge densities and nature of Fe-O1 and La-O1 bonds of lanthanum orthoferrite- type multiferroics are summarized in Table 3.6. The bond critical point charge density values in Table 3.6 reveal that the bond Fe-O1 is covalent and the bond La-O1 is also covalent for all the samples.

Table 3.5 *Bond angles and bond lengths in the Fe-O sub lattice of lanthanum orthoferrite-type multiferroics due to $< FeO_6 >$ octahedral twist*

Samples	Conc.	Bond angle (deg)	Bond length (Å)	Bond angle (deg)	Bond length (Å)
		Fe-O1-Fe	Fe-O1	Fe-O2-Fe	Fe-O2
	0.00	155.7997	2.0081	157.2900	1.9729
	0.03	155.5487	2.0084	157.3569	1.9568
LCFO	0.06	155.8081	2.0073	157.2904	1.9720
	0.09	154.9387	2.0074	156.9886	1.9730
	0.12	155.5344	2.0084	156.9582	1.9879
	0.00	157.8205	2.0020	157.1154	2.0699
LZFO	0.05	158.8469	1.9991	157.4606	2.0665
	0.15	158.6199	1.9990	157.1951	2.1016
	0.25	157.0968	2.0077	157.0600	2.0746
LAFO	0.05	160.8839	1.9914	157.8395	1.9946
	0.15	155.7094	2.0072	156.5067	1.9914
	0.25	157.0381	2.0000	156.9579	2.0006
LSFO	0.05	158.0220	1.9979	157.7021	2.0703
	0.10	161.8870	1.9858	158.1390	2.1305
	0.15	162.8048	1.9825	156.6920	2.1179
	0.20	164.2018	1.9647	158.9486	2.1693

LCFO- $La_{1-x}Ce_xFeO_3$ LZFO- $La_{1-x}Zn_xFeO_3$ LAFO- $La_{1-x}Al_xFeO_3$
LSFO- $La_{1-x}Sr_xFeO_3$
Conc. - concentration

Table 3.6 *Bond critical point charge density values and nature of Fe-O1 and La-O1 bonds of lanthanum orthoferrite-type multiferroics*

Sample	Conc (x)	Fe-O1		La-O1	
		Bond critical point charge density $(e/Å^3)$	Nature of bond with x	Bond critical point charge density $(e/Å^3)$	Nature of bond with x
LCFO	0.00	0.9251	covalent	0.9811	covalent
	0.03	0.7830	covalent	0.9162	covalent
	0.06	0.8340	More covalent	1.0741	More covalent
	0.09	0.6910	covalent	0.7983	covalent
	0.12	0.6830	covalent	0.8733	covalent
LZFO	0.00	0.6712	covalent	0.6872	covalent
	0.05	0.5553	covalent	0.5231	covalent
	0.15	0.5512	covalent	0.4326	covalent
	0.25	0.5929	More covalent	0.5671	More covalent
LAFO	0.05	0.8760	covalent	0.3175	covalent
	0.15	0.9465	covalent	0.4699	More covalent
	0.25	1.0150	More covalent	0.3249	covalent
LSFO	0.05	0.8693	More covalent	1.0203	More covalent
	0.10	0.8403	covalent	0.8540	covalent
	0.15	0.7835	covalent	0.7160	covalent
	0.20	0.7782	covalent	0.5281	covalent

LCFO - $La_{1-x}Ce_xFeO_3$
LZFO - $La_{1-x}Zn_xFeO_3$
LAFO - $La_{1-x}Al_xFeO_3$
LSFO - $La_{1-x}Sr_xFeO_3$
Conc. - concentration

3.7 Optical analysis of lanthanum orthoferrite-type multiferroics

The energy band gap of all the synthesized lanthanum orthoferrite-type multiferroics has been evaluated from UV-visible absorption data obtained using a UV- visible spectrophotometer. From the UV-Vis absorption spectra recorded in the wavelength range of 200 nm to 2000 nm, the energy band gap has been evaluated using the Wood and Tauc equation $\alpha h\upsilon = A(h\upsilon - E_g)^n$ [Wood and Tauc, 1972], where hv is

photon energy (E), α is the absorption coefficient, A is a material constant, E_g is energy band gap and n is 1/2 for direct band gap material LaFeO₃. The Tauc plot [Wood and Tauc, 1972] is drawn with photon energy along the x-axis and $(\alpha E)^2$ along the y-axis. Then the linear fit is applied to the plot and it is extrapolated to the *x*-axis, to estimate the energy band gap (E_g).

3.7.1 Ce-substituted lanthanum orthoferrites - $La_{1-x}Ce_xFeO_3$

The UV-visible absorption spectra of the synthesized $La_{1-x}Ce_xFeO_3$ (x=0.00, 0.03, 0.06, 0.09 & 0.12) multiferroics (the inset of Figure 2.41) shows the strong absorption peaks in the UV-region (~190-380 nm) centered at around 264 nm and the broad absorption bands in the visible region (~380-750 nm) centered at approximately 380 nm. The strong optical absorption peak in lanthanum orthoferrite-type multiferroics can be attributed to the electronic transition from the O $(2p) \rightarrow$ Fe $(3d)$ orbital [Li et al., 2000]. The inset of Figure 2.41 shows that the absorption peaks shift towards the lower wavelength side, which means the blue shift for the increasing substitution of Ce at the lattice site of La. The energy band gap (E_g) (Table 2.17) estimated from the Tauc plot [Wood and Tauc, 1972] of $La_{1-x}Ce_xFeO_3$ (x=0.00, 0.03 0.06, 0.09 & 0.12) (Figure 2.41) ranges from 2.22(0.36) eV to 2.41(0.11) eV. The E_g value of synthesized LaFeO₃ is close to the reported value [Köferstein et al., 2013]. The energy band gap in LaFeO₃ is due to the super-exchange interaction between Fe^{3+} electronic energy levels (unoccupied d-orbitals) through oxygen (p-orbitals) ion. For the synthesized $La_{1-x}Ce_xFeO_3$ multiferroics, the E_g values are found to increase with the increasing substitution of Ce at the La site. The reason for the increase in E_g is explained as follows: The substitution of Ce^{4+} at the La^{3+} site in LaFeO₃ leads to the decrease in the particle size in addition to the decrease in the number of Fe^{3+} ions with uncompensated spins. The decrease in the number of Fe^{3+} ions with unpaired electrons could lead to the reduction of the width of the available unoccupied d-orbitals of Fe^{3+} ions at the conduction band, which results in a shift of the conduction band minima to higher energies thereby increases the band gap. Furthermore, the decrease in particle size results in an increase in the number of uncompensated spins of Fe^{3+} ions and hence the significant direct energy band gap value for Ce-substituted samples.

3.7.1.1 Correlation between charge density and optical properties of $La_{1-x}Ce_xFeO_3$ multiferroics

The decrease in the charge density values along the Fe-O1 bond of Ce-substituted samples is attributed to the decrease in the number of Fe^{3+} ions with uncompensated spins, which are responsible for the increase in the energy band gap values.

3.7.2 Zn-substituted lanthanum orthoferrites - $La_{1-x}Zn_xFeO_3$

The UV-visible absorption spectra (the inset of Figure 2.42) of the $La_{1-x}Zn_xFeO_3$ (x=0.00, 0.05, 0.15 & 0.25) multiferroics show a strong absorption band at around 377nm in the visible region (~380−750 nm) and an infinitesimal blue shift of the absorption band for the increase in Zn concentration. The direct energy band gap values (Table 2.18) of $La_{1-x}Zn_xFeO_3$ (x=0.00, 0.05, 0.15 & 0.25) multiferroics estimated using the Tauc plot [Wood and Tauc, 1972] of $La_{1-x}Zn_xFeO_3$ (Figure 2.42) range from 2.025(0.21) eV to 2.097(0.13) eV. For the synthesized $La_{1-x}Zn_xFeO_3$ multiferroics, the energy band gap values are found to increase slightly with the increase in Zn concentration. This is due to the reason that when Zn substitution increases, the number of Fe^{3+} ions (as Fe^{3+} reduces to Fe^{2+} to maintain the charge neutrality in the system) with uncompensated spins (unpaired electrons), as well as particle size decreases. The decrease in the number of Fe^{3+} ions could lead to the reduction of the width of the available unoccupied d-orbitals of Fe^{3+} ionsat the conduction band and hence increases the band gap. Furthermore, the decrease in theparticle size results in the increase in the number of uncompensated spins of Fe^{3+} ions and hence the appreciable direct energy band gap value for Zn-substituted samples.

3.7.2.1 Correlation between charge density and optical properties of $La_{1-x}Zn_xFeO_3$ multiferroics

The decrease in the bond critical point charge density values along the Fe-O1bond of Zn-substituted samples when compared to the pristine sample is attributed to thedecrease in the number of Fe^{3+} ions which are responsible for the increase in their energyband gap values.

3.7.3 Al-substituted lanthanum orthoferrites - $La_{1-x}Al_xFeO_3$

The UV-visible absorption spectra (the inset of Figure 2.43) of the synthesized $La_{1-x}Al_xFeO_3$ (x=0.05, 0.15, 0.25 & 0.35) multiferroics show the strong optical absorption peaks at around 385 nm. The energy band gap values (Table 2.19) of $La_{1-x}Al_xFeO_3$ (x=0.00, 0.05, 0.15 & 0.25) multiferroics determined from the Tauc plot show that the E_g values range from 2.095(0.24) eV to 2.129(0.2) eV. The E_g values are found to decrease with the increase in Al concentration. The reason for this behavior of decrease in E_g is explained as follows: With the increasing substitution of Al, the particle size decreases. The decrease in the particle size can increase the number of uncompensated spins of Fe^{3+} ions that could lead to the increase of the width of the available unoccupied d-orbitals of Fe^{3+} ions at the conduction band and hence to the decrease of the band gap.

3.7.3.1 Correlation between charge density and optical properties of $La_{1-x}Al_xFeO_3$ multiferroics

The increase in the BCP charge density values along the bond Fe-O1 with the increase in substitution of Al is attributed to the increase in the number of uncompensated spins of Fe^{3+} ions which are responsible for the decrease in the energy band gap values.

3.7.4 Sr-substituted lanthanum orthoferrites - $La_{1-x}Sr_xFeO_3$

The UV-visible absorption spectra of the synthesized $La_{1-x}Sr_xFeO_3$ (x=0.05, 0.10, 0.15 & 0.20) multiferroics (the inset of Figure 2.44) depict that the strong absorption peaks centered at around 263 nm. For $La_{1-x}Sr_xFeO_3$ the energy band gap values (Table 2.20) range from 2.20(0.23) eV to 2.36(0.39) eV. There is a slight increase in E_g is observed with the increase in Sr concentration on the host lattice (Table 2.20). This is due to the reason that when Sr substitution increases, the number of Fe^{3+} ions (radius: 0.65 Å) [Shannon, 1976] in the host lattice decreases, while the number of Fe^{4+} ions (radius: 0.58 Å) [Shannon, 1976] increases. The decrease in the number of Fe^{3+} ions increases the band gap by decreasing the width of the available unoccupied d-orbitals of Fe^{3+} ions at the conduction band. Meanwhile, the decrease in the particle size leads to the increase in the number of uncompensated spins of Fe^{3+} ions which results in the appreciable energy band gap values.

3.7.4.1 Correlation between charge density and optical properties of $La_{1-x}Sr_xFeO_3$ multiferroics

The decrease in the charge density values along the primary bond Fe-O1 with the increase of Sr concentration is attributed to the decrease in the number of Fe^{3+} ions which are responsible for the increase in the energy band gap values.

The energy band gap values of the synthesized lanthanum orthoferrite-type multiferroics estimated using UV-visible spectra estimated from and the Tauc plot are tabulated in Table 3.7.

Table 3.7 *Energy band gaps of lanthanum orthoferrite-type multiferroics using UV-visible spectra*

Samples	Conc.(x)	Band gap (eV)
LCFO	0.00	2.22(0.36)
	0.03	2.23(0.56)
	0.06	2.26(0.19)
	0.09	2.34(0.13)
	0.12	2.41(0.11)
LZFO	0.00	2.025(0.21)
	0.10	2.048(0.2)
	0.15	2.074(0.27)
	0.25	2.097(0.13)
LAFO	0.05	2.129(0.2)
	0.15	2.112(0.19)
	0.25	2.095(0.24)
LSFO	0.05	2.20(0.23)
	0.10	2.24(0.34)
	0.15	2.30(0.2)
	0.20	2.36(0.39)

LCFO - $La_{1-x}Ce_xFeO_3$
LZFO - $La_{1-x}Zn_xFeO_3$
LAFO - $La_{1-x}Al_xFeO_3$
LSFO - $La_{1-x}Sr_xFeO_3$
Conc. - concentration

3.8 Magnetic properties of lanthanum orthoferrite-type multiferroics

The magnetic properties of the synthesized lanthanum orthoferrite-type multiferroics have been analyzed by extracting the data from the M-H curves recorded at room temperature using a vibrating sample magnetometer.

3.8.1 Ce-substituted lanthanum orthoferrites - $La_{1-x}Ce_xFeO_3$

Lanthanum orthoferrite ($LaFeO_3$) is a G-type antiferromagnetic and insulating material with a high Néel temperature (T_N) of 740°C [Hearne and Pasternak, 1995].The antiferromagnetic ordering in this system is due to the super-exchange interactions between the magnetic (iron) atoms through oxygen atoms, which force all the Fe^{3+} spinsto be antiferromagnetically aligned between the two sub-lattices. In general, the La^{3+} ionis inert to magnetic interactions because it has paired valence electrons. Therefore, the magnetic moments of iron ion and the antisymmetric exchange interaction between the iron ion (Fe^{3+}) and the rare earth ion (La^{3+}) are the primary sources of the magnetic properties of $LaFeO_3$.

The room temperature magnetic hysteresis (M-H) curves of the synthesized $La_{1-x}Ce_xFeO_3$ (x=0.00, 0.03, 0.06, 0.09 & 0.12) multiferroics (Figure 2.45) and the enlarged M-H curves (inset of Figure 2.45) show that all the samples exhibit weak ferromagnetic (FM) behavior with appreciable magnetization. The weak ferromagnetism in $LaFeO_3$ is promoted by the spin canting of antiferromagnetically ordered Fe moments that arise as a result of Fe^{3+}–O^{2-}–Fe^{3+} super-exchange interaction [Azab et al., 2015, Qi et al., 2003, Shen et al., 2009, Shika et al., 2015].

Figure 2.45 also shows unsaturated magnetic hysteresis curves, which reflect the existence of both ferro and anti-ferromagnetic ordering in all the samples. Furthermore, from Figure 2.45, it is seen that the M-H curves of all the samples are biased in the presence of an applied field of about 15000 G, showing a shift in the negative direction, which is attributed to the negative exchange bias (EB) effect. The negative EB is caused by the exchange coupling between the spins of ferromagnetic (FM) and anti-ferromagnetic (AFM) domains present in the samples. Thus, the existence of both FM and AFM domains results in asymmetric magnetization hysteresis. In addition, the interaction among the FM and AFM domains can strongly control the reversal mechanism of the magnetization, modifying nucleation and annihilation of the vortex [Dantas et al., 2005, Berkowitz and Kentaro, 1999]. The asymmetry in the remanent magnetization is attributed to the exchange bias anisotropy [Berkowitz and Kentaro, 1999]. The coercive

field and exchange bias field of $La_{1-x}Ce_xFeO_3$ were determined from the M-H curves, using the relations,

$$H_C = \frac{(H_{C+} - H_{C-})}{2} \;\ldots\ldots\ldots \tag{3.1}$$

$$H_{EB} = -\frac{(H_{C+} + H_{C-})}{2} \;\ldots\ldots\ldots \tag{3.2}$$

where, H_{C+} and H_{C-} are the intercepts of the magnetization (M) on the +ve and –ve side of the applied magnetic field (H) axis.

Magnetic parameters (Table 2.21) show that the substitution of Ce in $LaFeO_3$ enhances its magnetic properties. The reasons for the enhanced magnetic properties arethe increase in the substitution of Ce leads to the increase of uncompensated Fe^{3+} spins as a result of the decrease in particle size and the change in the Fe–O–Fe canting angle as a result of antisymmetric exchange interaction. In the Ce-substituted samples, the charge compensation that takes place due to the presence of Ce^{4+} ion, which converts Fe^{3+} ions into Fe^{2+} ions [Shikha et al., 2015] also influences magnetic properties. The values of H_C and H_{EB} are found to be high as 1150 G and 1490 G respectively, for x=0.06, compared to other Ce-substituted samples, which are due to the fact that the probability of formation of AFM domains is more than that of FM domains in this level of substitution. The high value of H_C for this sample (x=0.06) is also attributed to the increase of magnetocrystalline anisotropy, which arises as a result of the increase of the AFM spins [Giannakas et al., 2006] and the high value of H_{EB} is due to the additional exchange bias field produced in addition to the field that is already present in the AFM domains. The additional H_{EB} in this sample is originated through the interfacial interaction between the spins at the interface of AFM and FM domains, which tries to align ferromagnetically the FM spins with the AFM spins at the interface [Ahmed et al., 2015]. The low value of M_s (M_s = 0.167 emu/g) for this sample (x=0.06) confirms that the probability of formation of the FM domains is less in this substitution level.

3.8.1.1 Correlation between charge density and magnetic properties of $La_{1-x}Ce_xFeO_3$ multiferroics

The magnetic properties of $La_{1-x}Ce_xFeO_3$ have been analyzed through chargedensities. The high values of charge density along the bonds Fe-O1 and La-O1 of the pure sample (x=0.00) confirm the presence of Fe^{3+} spins and antisymmetric exchange interaction between La^{3+} and Fe^{3+} ions. The decrease in the charge density values along the Fe-O1 bond of the Ce-substituted samples is attributed to the introduction of Fe^{2+} ions

in the host lattice which is caused by the charge compensation of substituted Ce^{4+} ions [Giannakas et al., 2006]. The introduction of Fe^{2+} ions in the crystal lattice weakens the antisymmetric exchange interactions thereby decreases the magnetic behavior. But the presence of uncompensated Fe^{3+} spins, which are caused as a result of the decrease in particle size leads to the strengthening of the antisymmetric exchange interaction, and hence the enhancement in the magnetic behavior of Ce-substituted samples. The significant values of BCP charge density along the bonds Fe-O1 and La-O1 for the Ce-substituted samples are good evidence for the presence of Fe^{3+} spins and strengthened antisymmetric exchange interaction. The high values of BCP charge density along the bonds Fe-O1 and the La-O1 for the sample with moderate Ce concentration x=0.06 is due to the electronic charge transfer from outer 5p shell to the empty 5d and 4f subshells in some Ce^{4+} ions [Bhushan et al., 2012] in addition to the Fe^{3+} spins in this substitution level. For x=0.09, saturation magnetization (M_s) and remanent magnetization (M_r) are found to be higher than those observed for x=0.12, which is due to the presence of more Fe^{3+} spins. The relatively high value of charge density along the Fe-O1 bond of the sample with x=0.09 is good evidence for the presence of more Fe^{3+} spins which are responsible for its high saturation and remanent magnetizations.

3.8.2   Zn-substituted lanthanum orthoferrites - $La_{1-x}Zn_xFeO_3$

The room temperature magnetic hysteresis (M-H) curves of the synthesized $La_{1-x}Zn_xFeO_3$ (x=0.00, 0.05, 0.15 & 0.25) multiferroics (Figure 2.46) and the enlarged M-H curves (the inset of Figure 2.46) show narrow-width hysteresis curves, which reflect that the prepared samples are soft in nature with weak ferromagnetic (FM) behavior. The ferromagnetic ordering in $La_{1-x}Zn_xFeO_3$ is promoted by the uncompensated spins of Fe^{3+} ions, which causes the spin canting of Fe moments by super-exchange interaction between Fe metal atoms via oxygen atom [Azab et al., 2015, Qi et al., 2003, Shen et al., 2009, Shika et al., 2015]. Figure 2.46 also illustrates the unsaturated M-H curves at the high magnetic field of about 15000 G, which reflects the co-existence of ferro and antiferromagnetism in all the samples. From the M-H curves, it is also noticed that the hysteresis curves of all the samples are biased showing a shift in the negative direction which may be attributed to the negative exchange bias (EB) effect caused due to the exchange coupling between the spins of ferromagnetic and antiferromagnetic domains present in the samples. Thus, the co-existence of ferromagnetic (FM) and antiferromagnetic (AFM) domains results in asymmetric magnetization hysteresis curves.

The asymmetry in the remanent magnetization is attributed to the exchange bias anisotropy [Berkowitz and Kentaro, 1999].

From the hysteresis curves of $La_{1-x}Zn_xFeO_3$, the coercive field and exchange bias field are estimated using the equations (3.1) and (3.2) respectively. The magnetic parameters such as saturation magnetization (M_s) and remanent magnetization (M_r), coercive field (H_C) and exchange bias field (H_{EB}) have been extracted from the M-H curves of $La_{1-x}Zn_xFeO_3$ and are listed in Table 2.22. From Table 2.22, it is evident that the values of M_r, H_C, and H_{EB} decrease, whereas the saturation magnetization (M_s) increases for Zn-substituted samples as compared to pure $LaFeO_3$, which is due to the probability of formation of FM domains is more than that of AFM domains. The values of H_C and H_{EB} for x=0.00 are 295 G and 570 G respectively. The pure sample shows higher H_C and H_{EB} than Zn-substituted samples, which is due to the probability of the formation of AFM domains is more than that of FM domains in this sample. For this (x=0.00) sample, the value of M_s is found to be low as

0.18 emu/g, which confirms that the formation of the AFM domains is more in this system. The higher value of H_C for this sample (x=0.00) is due to the high magnetocrystalline anisotropy that arises as a result of more AFM spins [Giannakas et al., 2006]. Likewise, the higher value of H_{EB} is due to the additional exchange bias field caused by the interfacial interaction between the spins at the interface of AFM and FM domains in addition to the field that is already present in the AFM domain [Ahmed et al., 2015]. Among the Zn-substituted samples, for x=0.25, the value of saturation magnetization (M_s) is found to be high as 0.53 emu/g, which is due to the formation of more FM domains caused by the presence of more uncompensated spins of Fe^{3+} ions, which arises due to decrease in particle size. Furthermore, for x=0.25, the value of H_{EB} is found to be low as 139 G. The low value of H_{EB} is attributed to the decrease of magnetocrystalline anisotropy that arises as a result of the increase in the FM spins. Thus the reasons for the ferromagnetic property with enhanced M_s values observed in Zn-substituted samples are the increase in the number of uncompensated Fe^{3+} spins as a result of the decrease in particle size and also the change in the Fe–O–Fe canting angle as a result of antisymmetric exchange interaction.

3.8.2.1 Correlation between charge density and magnetic properties of $La_{1-x}Zn_xFeO_3$ multiferroics

The magnetic properties of $La_{1-x}Zn_xFeO_3$ multiferroics have been analyzed through charge density distribution. The high values of BCP charge density along the bonds Fe-O1 and La-O1 of x=0.00 indicate the presence of Fe^{3+} spins and antisymmetric exchange interaction between La^{3+} and Fe^{3+} ions respectively. The decrease in the charge density values along the bonds Fe-O1 and La-O1 of Zn-substituted samples may be attributed to the decrease in the number of Fe^{3+} ions to maintain charge neutrality in the host lattice [Manzoor and Husain, 2018]. The decrease in the number of Fe^{3+} ions is expected to cause the weakening of antisymmetric exchange interaction and hence to the decrease in magnetic behavior. But the presence of uncompensated Fe^{3+} spins that arises as a result of the decrease in particle size promotes strengthening of the antisymmetric exchange interaction and hence the enhancement in the magnetic behavior of Zn-substituted samples. The appreciable values of charge density along the bonds Fe-O1 and La-O1 for the Zn-substituted samples are good evidence for the presence of uncompensated Fe^{3+} spins and strengthened antisymmetric exchange interaction respectively. For x=0.25, the value of maximum magnetization (M_s) is found to be relatively higher than that observed for other Zn concentrations, which is due to the presence of more uncompensated spins of Fe^{3+} ions. The relatively high values of charge density along the bonds Fe-O1 and La-O1 for x=0.25 are good evidence for the presence of more uncompensated Fe^{3+} spins and strengthened antisymmetric exchange interaction respectively, which are responsible for its enhanced magnetic behavior.

3.8.3   Al-substituted lanthanum orthoferrites - $La_{1-x}Al_xFeO_3$

The room temperature magnetic hysteresis (M-H) curves of the synthesized $La_{1-x}Al_xFeO_3$ (x=0.05, 0.15 & 0.25) multiferroics (Figure 2.47) and the enlarged M-H curves of $La_{1-x}Al_xFeO_3$ (x=0.05 & x=0.15) (inset of Figure 2.47) reveals that all the samples behave like a weak ferromagnet. The weak ferromagnetism is originated by canting of antiferromagnetically ordered metal ion (Fe^{3+}) spins that arises as a result of Fe^{3+}–O^{2-}–Fe^{3+} super-exchange interaction [Azab et al., 2015, Qi et al., 2003, Shen et al., 2009, Shika et al., 2015]. The antisymmetric exchange interaction between the magnetic (Fe^{3+}) ions and the rare-earth (La^{3+}) ions also plays a vital role in modifying the magnetic properties of $LaFeO_3$ [Qi et al., 2003, Shen et al., 2009, Shikha et al., 2015, Azab et al., 2015]. The magnetic hysteresis curves also illustrate the jump-like magnetization at zero magnetic fields termed Barkhausen jumps [García et al., 2004]. This jump-like magnetization in the synthesized $La_{1-x}Al_xFeO_3$ multiferroics is due to the nucleation and orientation of ferromagnetic domains in the stressed region, which was created due to

crystal inhomogeneities. The magnetic parameters such as saturation magnetization (M_s), remanent magnetization (M_r) and coercive field (H_C) of $La_{1-x}Al_xFeO_3$ have been extracted from their hysteresis loop and have been provided in Table 2.23.

Table 2.23 indicates that the values of M_s, M_r and H_C increase with the increasing substitution of Al. The increased magnetic parameters (M_s and M_r) can be explained by the fact that the increasing substitution of Al^{3+} ions at the La^{3+} site in $LaFeO_3$ leads to a decrease in particle size, which increases uncompensated spins of Fe^{3+} ions [Kodama et al., 1997, Winkler et al., 2005]. The increase of uncompensated spins of Fe^{3+} ions results in significant antisymmetric exchange interaction with the La^{3+} ions, thus exhibiting enhanced magnetic properties. The increase in coercivity (H_C) of $La_{1-x}Al_xFeO_3$ with the increase in the substitution of Al is attributed to the increase in the magnetocrystalline anisotropy which arises due to the decrease in particle size [Ahmed et al., 2015].

3.8.3.1 Correlation between charge density and magnetic properties of $La_{1-x}Al_xFeO_3$ multiferroics

The magnetic properties of $La_{1-x}Al_xFeO_3$ have been analyzed through charge density values. The increase in the values of BCP charge density along the bond Fe-O1 with the increase in Al concentration is good evidence for the increase in the number of uncompensated spins of Fe^{3+} ions and the significant values of BCP charge density along the bond La-O1 are good evidence for the significant antisymmetric exchange interaction between the Fe^{3+} and La^{3+} ions which are responsible for the enhanced magnetic behavior of the synthesized $La_{1-x}Al_xFeO_3$ multiferroics.

3.8.4 Sr-substituted lanthanum orthoferrites - $La_{1-x}Sr_xFeO_3$

The room temperature magnetic hysteresis (M-H) curves of the synthesized $La_{1-x}Sr_xFeO_3$ (x=0.05, 0.10, 0.15 & 0.20) multiferroics (Figure 2.48) show that all the samples exhibit ferromagnetic (FM) behavior with appreciable magnetization and coercivity. The weak ferromagnetism can be explained by the canting of antiferromagnetically ordered metal ion (Fe^{3+}) spins that takes place as a result of super- exchange interaction between Fe^{3+} ions via O^{2-} ions [Azab et al., 2015, Qi et al., 2003,

Shen et al., 2009, Shika et al., 2015]. Various magnetic parameters such as saturation magnetization (M_s), remanent magnetization (M_r) and coercive field (H_C) of $La_{1-x}Sr_xFeO_3$ extracted from the hysteresis curves are presented in Table 2.24. From Table 2.24, it is evident that for the sample with Sr concentration x=0.05, the values of magnetic parameters such as M_s, M_r and H_C are found to be high as 2.4 emu/g, 1.12 emu/g and 1616 G respectively. The high M_s, M_r and H_C values for x=0.05, are due to the presence of more Fe^{3+} ions with uncompensated spins, which strengthens the antisymmetric exchange interaction between the Fe^{3+} ions and the La^{3+} ions. The value of magnetic parameters such as M_s, M_r, and H_C are found to decrease with the increase in Sr concentration. The decrease in the magnetic behavior is due to the decrease in the number of Fe^{3+} ions, which weakens the antisymmetric exchange interaction.

3.8.4.1 Correlation between charge density and magnetic properties of $La_{1-x}Sr_xFeO_3$ multiferroics

The magnetic properties of $La_{1-x}Sr_xFeO_3$ have been analyzed through charge

density studies. The high values of charge density along the bonds Fe-O1 and La-O1 for x=0.05 are good evidence for the presence of the higher number of Fe^{3+} ions with uncompensated spins and strengthened antisymmetric exchange interaction respectively, which are responsible for its enhanced magnetic behavior. The decrease in the charge density values along the bonds Fe-O1 and La-O1 with the increase in Sr concentration is attributed to the decrease in the number of Fe^{3+} ions and weakened antisymmetric exchange interaction respectively, which are responsible for the decreased magnetic behavior.

The magnetic parameters of the synthesized lanthanum orthoferrite-type multiferroics extracted from the magnetic hysteresis curves have been tabulated in Table 3.8.

Table 3.8 *Magnetic parameters of lanthanum orthoferrite-type multiferroics*

Samples	Conc. (x)	M_s (emu/g)	M_r (emu/g)	H_{C+} (G)	H_{C-} (G)	H_C (G)	H_{EB} (G)
0.00		0.141	0.036	123	1848	863	986
0.03		0.195	0.040	101	1911	905	1006
LCFO	0.06	0.167	0.052	340	2639	1150	1490
0.09		0.256	0.053	165	2314	1075	1240
0.12		0.232	0.045	229	2081	926	1155
0.00		0.18	0.022	275	865	295	570
LZFO	0.05	0.44	0.072	267	325	29	296
0.15		0.28	0.014	115	261	73	188
0.25		0.53	0.017	80	198	59	139
LAFO	0.05						
		0.20	0.03	-	-	1044	
0.15		0.44	0.17	-	-	1992	
0.25		3.73	2.02	-	-	3031	
	0.05	2.4	1.12	-	-	1616	
LSFO							
0.10		1.65	0.74	-	-	1210	
0.15		1.46	0.59	-	-	1057	
0.20		1.7	0.66			917	

M_s - Saturation magnetization, M_r - Remanent magnetization , H_{C+}, H_{C-} - Intercepts of magnetization on the +ve and –ve side of field axis, H_C - Coercivefield, H_{EB} - Exchange bias field.
LCFO - $La_{1-x}Ce_xFeO_3$
LZFO - $La_{1-x}Zn_xFeO_3$
LAFO - $La_{1-x}Al_xFeO_3$
LSFO - $La_{1-x}Sr_xFeO_3$
Conc. - concentration

3.9 Dielectric properties of lanthanum orthoferrite-type multiferroics

The dielectric characterization of all the synthesized multiferroics has been carriedout in the wide frequency range from 10 Hz to 5 MHz at room temperature using an impedance analyzer.

3.9.1   Ce-substituted lanthanum orthoferrites - $La_{1-x}Ce_xFeO_3$

The frequency (f) dependent dielectric constant (ε') plot of $La_{1-x}Ce_xFeO_3$ multiferroics (Figure 2.49) and the inset, which shows the (ε'-f) plot drawn in the frequency range of 10 Hz to 5 MHz depict that all the samples exhibit the normal dielectric dispersion. The observed dielectric response is attributed to space charge polarization as described by Maxwell, Wagner and Koop [Maxwell, 1973, Wagner, 1993, Koops, 1951]. In $LaFeO_3$, the high ε' at the low-frequency region, is mainly due to the response of $Fe^{2+} \leftrightarrow Fe^{3+}$ dipoles to the applied field, and the almost constant ε' at the very high-frequency region is due to the inability of $Fe^{2+} \leftrightarrow Fe^{3+}$ dipoles to respond to the applied field. The frequency (f) dependent dielectric loss (tan δ) plot of $La_{1-x}Ce_xFeO_3$ (x=0.00, 0.03, 0.06, 0.09 & 0.12) (Figure 2.50) measured in the frequency range of 10 Hz to 5 MHz shows the decreasing trend of tan δ with increasing frequency and it becomes nearly constant at high frequencies. The ac conductivity of $La_{1-x}Ce_xFeO_3$ was estimated from the dielectric data using the formula [Chougule and Chougule, 2007]

$$\sigma_{ac} = 2\pi\varepsilon_0\varepsilon' f \tan \delta \qquad\qquad \dots\dots\dots (3.3)$$

where. ε_0 is the permittivity of free space (ε_0 = 8.854 x 10^{-12} f/m), ε' is the dielectric constant of the sample, f is the frequency of the external field and tan δ is the dielectric loss factor (dielectric loss tangent). The frequency (f) dependent ac conductivity (σ_{ac}) plot of $La_{1-x}Ce_xFeO_3$ (x=0.00, 0.03, 0.06, 0.09 & 0.12) (Figure 2.51) and the inset, which shows the (σ_{ac}-log f) plot of $LaFeO_3$ depict that the ac conductivity remains constant in the low-frequency region and increases with the increase in applied frequency for all the samples. The increasing ac conductivity with increasing frequency in $LaFeO_3$ is attributed to the effect of the applied alternating electric field that promotes hopping of charge carriers between $Fe^{2+} \leftrightarrow Fe^{3+}$ ions (n-type charge carriers) and $La^{3+} \leftrightarrow La^{4+}$ ions (p-type charge carriers). The dielectric constant (ε'), dielectric loss (tan δ) and ac conductivity (σ_{ac}) obtained from the dielectric measurements of $La_{1-x}Ce_xFeO_3$ are given in Table 2.25.

From Table 2.25, it is evident that ε' for pristine $LaFeO_3$ is 27998, which is much higher than that for Ce-substituted samples. The dielectric loss (tan δ) value for the pristine sample is observed as 9.4 and for the Ce-substituted samples, it ranges from 3.5 to 41. The ac conductivity for the pristine sample is 0.3389 $\Omega^{-1}m^{-1}$, which is higher than that for Ce-substituted samples. The decrease in dielectric property and ac conductivity in Ce-substituted samples is due to reason that the incorporation of Ce^{4+} ions in the $LaFeO_3$ system reduces the Fe^{3+} ions in the system that are responsible for both space charge polarization and hopping of charge carriers between the localized states. Among the Ce-substituted samples for x=0.06, the values of ε' and ac conductivity are found to be relatively high as 5070 and 0.008 $\Omega^{-1}m^{-1}$, which is due to the presence of more cations (Fe^{3+} and La^{3+} ions) with n-type and p-type charge carriers.

3.9.1.1 Correlation between charge density and dielectric properties of $La_{1-x}Ce_xFeO_3$ multiferroics

The decrease in the BCP charge density values along the bonds Fe-O1 and La-O1 is good evidence for the reduction of charge carriers (Fe^{3+} ions and La^{3+} ions) in the Ce-substituted samples which are responsible for the decrease in their dielectric property and ac conductivity. Furthermore, the high values of BCP charge density along the bonds Fe-O1 and La-O1 for x= 0.06, confirm the presence of more charge carriers which are responsible for its high dielectric constant and ac conductivity.

3.9.2 Zn-substituted lanthanum orthoferrites - $La_{1-x}Zn_xFeO_3$

Figure 2.52 shows the frequency (f) dependent dielectric constant (ε') plot of $La_{1-x}Zn_xFeO_3$ (x=0.00, 0.05, 0.15 & 0.25) multiferroics and the inset shows the frequency

(f) dependent dielectric constant (ε') measured in the frequency limit of 10 Hz to 5 MHz. From Figure 2.52, it is seen that the synthesized $La_{1-x}Zn_xFeO_3$ exhibits normal dielectric behavior, which can be explained in terms of Maxwell-Wagner type polarization and Koop's theory [Maxwell, 1973, Wagner, 1993, Koops, 1951]. The value of the dielectric constant (ε') is much higher at lower frequencies due to the response of $Fe^{2+} \leftrightarrow Fe^{3+}$ dipoles to the external field and its value decreases with the increasing frequency of the applied electric field. At higher frequencies, ε' shows no change due to the inability of $Fe^{2+} \leftrightarrow Fe^{3+}$ dipoles to respond to the external field. The frequency (f) dependent dielectric loss (tan δ) plot of $La_{1-x}Zn_xFeO_3$ (x=0.00, 0.05, 0.15 & 0.25) multiferroics (Figure 2.53) illustrates that tan δ decreases with increasing frequency from 10 Hz to5 MHz. The frequency (f) dependent ac conductivity (σ_{ac}) plot of $La_{1-x}Zn_xFeO_3$ (x=0.00, 0.05, 0.15 & 0.25) multiferroics (Figure 2.54) illustrates that the conductivity of all the samples remains constant in the low-frequency region and increases with the increase in applied frequency. In $La_{1-x}Zn_xFeO_3$, the increasing trend of the ac conductivity with the increasing applied field frequency is attributed to the effect of the applied field, which promotes the hopping of charge carriers between $Fe^{2+} \leftrightarrow Fe^{3+}$ ions (n-type charge carriers) and $La^{3+} \leftrightarrow La^{4+}$ ions (p-type charge carriers). The values of dielectric constant (ε'), dielectric loss (tan δ) and ac conductivity (σ_{ac}) of $La_{1-x}Zn_xFeO_3$ are tabulated in Table 2.26.

Table 2.26, shows that for pure LaFeO$_3$, the dielectric constant (ε') is 9.55 x 10^2 with a high dielectric loss (tan δ) of about 5.9 and ac conductivity is relatively low as 0.0028 $\Omega^{-1}m^{-1}$. Among the Zn-substituted samples for x=0.25, the dielectric constant is found to be appreciably high as 1.37 x 10^3, with a low dielectric loss and its ac conductivity is found to be relatively high as $3.6 \times 10^{-3}\ \Omega^{-1}m^{-1}$. The high values of ε' and σ_{ac} for x=0.25 are due to the presence of more uncompensated spins of Fe^{3+} ions that are responsible for both space charge polarization and hopping of charge carriers between the localized states $Fe^{2+} \leftrightarrow Fe^{3+}$ (n-type charge carriers) and $La^{3+} \leftrightarrow La^{4+}$ (p-type charge carriers). For pure $LaFeO_3$ multiferroic, the dielectric constant is reported as ε' < 10^3 [Idrees et al., 2011]. The value of ε' for pure $LaFeO_3$ obtained in this work is about 9.5×10^2 (i.e. ε' < 10^3), which is in agreement with the reported value [Idrees et al., 2011].

3.9.2.1 Correlation between charge density and dielectric properties of $La_{1-x}Zn_xFeO_3$ multiferroics

The high BCP charge density values observed along the bonds Fe-O1 and La-O1 for x=0.25 are good evidence for the presence of more uncompensated spins of Fe^{3+} ions and charge carriers that are responsible for its high ε' and σ_{ac}.

3.9.3 Al-substituted lanthanum orthoferrites - $La_{1-x}Al_xFeO_3$

The frequency (f) dependent dielectric constant (ε') plot of $La_{1-x}Al_xFeO_3$ (x=0.05, 0.15 & 0.25) multiferroics (Figure 2.55) shows normal dielectric dispersion, which is mainly attributed to the dipoles ($Fe^{2+} \leftrightarrow Fe^{3+}$) constituted by cations with different valencestates and space charge polarization as explained by Maxwell–Wagner [Maxwell, 1973, Wagner, 1993] and Koops [Koops, 1951]. At low frequencies, the value of dielectric constant is high because the magnetic iron dipoles ($Fe^{2+} \leftrightarrow Fe^{3+}$) have the potential to follow the frequency of the applied electric field, whereas, in the high-frequency region, the value of dielectric constant decreases and almost remains constant due to the inability of the $Fe^{2+} \leftrightarrow Fe^{3+}$ dipoles to respond to the frequency of the applied electric field.

The frequency-dependent dielectric loss plot of $La_{1-x}Al_xFeO_3$ (x=0.05, 0.15 & 0.25) multiferroics (Figure 2.56) illustrates that the dielectric loss follows the same trend as the dielectric constant over the whole frequency range of 10 Hz to 5 MHz. The plot of frequency-dependent ac conductivity (Figure 2.57) depicts dispersion for the applied frequency. Figure 2.57 shows that conductivity in the synthesized samples is due to the small polaron hopping and it is almost constant at low frequencies but increases rapidly after a certain frequency. The reason for the constant conductivity at low frequencies is due to the inertia of charge carriers, whereas the increase of conductivity at high frequencies is due to the hopping of charge carriers between $Fe^{2+} \leftrightarrow Fe^{3+}$ ions (*n*-type charge carriers) and $La^{3+} \leftrightarrow La^{4+}$ ions (p-type charge carriers), which are promoted by the applied electric field.

The dielectric constant (ε'), dielectric loss (tan δ) and ac conductivity (σ_{ac}) of $La_{1-x}Al_xFeO_3$ obtained from the dielectric measurements are given in Table 2.27. Table 2.27 shows that for the prepared $La_{1-x}Al_xFeO_3$, the dielectric constant (ε') and dielectric loss (tan δ) lies in the range of 213 to 771 and 0.77 to 2.97 respectively. Again Table 2.27 indicates that the ac conductivity (σ_{ac}) of $La_{1-x}Al_xFeO_3$ multiferroics ranges from 1.07×10^{-3} $\Omega^{-1}m^{-1}$ to 1.51×10^{-3} $\Omega^{-1}m^{-1}$. For x=0.25, the value of dielectric constant is found to be high as 771, due to high space charge polarization caused by the rotational displacement of more $Fe^{2+} \leftrightarrow Fe^{3+}$ dipoles in addition to the local displacement of $La^{2+} \leftrightarrow La^{3+}$ dipoles. The more $Fe^{2+} \leftrightarrow Fe^{3+}$ dipoles in this sample (x=0.25) are due to the increase in the number of uncompensated spins of Fe^{3+} ions, which occurred as a result of the decrease in particle size with the increasing substitution of Al concentration. From Table 2.27, it is also seen that for x=0.15, ac conductivity (σ_{ac}) is found to be slightly high as 1.51×10^{-3} $\Omega^{-1}m^{-1}$, which is attributed to the contribution of p-type charge carriers to the electrical conduction is more in this substitution level.

3.9.3.1 Correlation between charge density and dielectric properties of $La_{1-x}Al_xFeO_3$ multiferroics

The high BCP charge density along the Fe-O1 bond and the appreciable value of charge density along the La-O1 bond for x=0.25 are good evidence for the increase in the number of uncompensated spins of Fe^{3+} ions and hence more $Fe^{2+} \leftrightarrow Fe^{3+}$ dipoles, and the presence of $La^{2+} \leftrightarrow La^{3+}$ dipoles respectively, which are responsible for its high space charge polarization. Also, the high BCP charge density along the La-O1 bond of x=0.15 system is good evidence for the presence of more p-type charge carriers that are responsible for its high conductivity.

3.9.4   Sr-substituted lanthanum orthoferrites - $La_{1-x}Sr_xFeO_3$

Figure 2.58 depicts the frequency (f) dependent dielectric constant (ε') plot of $La_{1-x}Sr_xFeO_3$ (x=0.05, 0.10, 0.15 & 0.20) multiferroics and the inset shows the frequency dependent dielectric constant (ε') plot of $La_{1-x}Sr_xFeO_3$ (x=0.10, 0.15 & 0.20). Figure 2.58 shows the prepared $La_{1-x}Sr_xFeO_3$ multiferroics exhibit normal dielectric dispersion due to Maxwell-Wagner type polarization [Maxwell, 1973, Wagner, 1993], which is also in agreement with Koop's phenomenological theory [Koops, 1951]. In the (ε'-f) plot, the high ε' at the low-frequency region is due to the response of $Fe^{3+} \leftrightarrow Fe^{4+}$ dipoles to the applied field, and the almost constant ε' at the very high-frequency region is due to the inability of $Fe^{3+} \leftrightarrow Fe^{4+}$ dipoles to respond to the applied field. The frequency (f) dependent dielectric loss (tan δ) plot of $La_{1-x}Sr_xFeO_3$ (x=0.05, 0.10, 0.15 & 0.20) multiferroics (Figure 2.59) illustrates that the value of tan δ decreases with increasing frequency and become nearly constant at higher frequencies. The inset (Figure 2.59) shows (tan δ-f) plot of $La_{1-x}Sr_xFeO_3$ (x=0.05, 0.10 & 0.15). The frequency (f) dependent ac conductivity (σ_{ac}) plot of $La_{1-x}Sr_xFeO_3$ (x=0.05, 0.10, 0.15 & 0.20) multiferroics (Figure 2.60) and the enlarged (σ_{ac}-log f) plot (inset Figure

162

2.60) shows the conductivity remains constant in the low-frequency region and increases with the increase in applied frequency. The increase in σ_{ac} with the increase in applied frequency is attributed to the effect of the applied alternating electric field that promotes the hopping of charge carriers among $Fe^{3+} \leftrightarrow Fe^{4+}$ ions (n-type charge carriers) and $La^{3+} \leftrightarrow La^{4+}$ ions (p-type charge carriers). The values of dielectric constant (ε'), dielectric loss (tan δ) and ac conductivity (σ_{ac}) of $La_{1-x}Sr_xFeO_3$ are presented in Table 2.28. Table 2.28 shows the dielectric constantand conductivity decreases with increasing Sr concentration, which may be due to the inclusion of Sr^{2+} ions decreases the Fe^{3+} ions and La^{3+} ions that are responsible for space charge polarization and hopping of the charge carriers between the localized states. From Table 2.28, it is evident that for x=0.05, the ε' and σ_{ac} are found to be relatively high as 2.32 x 10^5 and 0.2081 $\Omega^{-1}m^{-1}$. The giant dielectric constant and high conductivity for the sample with x=0.05 is attributed to the presence of more cations (Fe^{3+} and La^{3+} ions)with n-type and p-type charge carriers, which can hop among the $Fe^{3+} \leftrightarrow Fe^{4+}$ ions and $La^{3+} \leftrightarrow La^{4+}$ ions. From Table 2.28, it is also evident that the dielectric loss (tan δ) lies in the range of 5 to 955. Since the dielectric constant for x=0.05, is relatively high, the dielectric loss factor corresponding to this concentration can be considered as low.

3.9.4.1 Correlation between charge density and dielectric properties of $La_{1-x}Sr_xFeO_3$ multiferroics

The decrease in the charge density values along the Fe-O1 and La-O1 bonds with the increase in Sr concentration is good evidence for the reduction of Fe^{3+} and La^{3+} cations, which are responsible for the decrease in dielectric constant and ac conductivity. The high values of charge density along the bonds Fe-O1 and La-O1 for x=0.05, are good evidence for the presence of more charge carriers that are responsible for its high dielectric constant and ac conductivity.

The dielectric parameters and ac conductivity of the synthesized lanthanum orthoferrite-type multiferroics estimated from the dielectric measurements have beensummarized in Table 3.9.

Table 3.9 *Dielectric parameters of lanthanum orthoferrite-type multiferroics*

Samples	Conc. (x)	$a\varepsilon'$	$a\tan\delta$	$b\sigma ac$ $(\Omega\text{-1m-1})$
LCFO				
	0.00	27998	9.4	0.3389
	0.03	811	41	0.0014
	0.06	5070	5.7	0.0080
	0.09	3158	3.5	0.0009
	0.12	1032	13.9	0.0007
LZFO				
	0,00	955	5.9	0.0028
	0.05	397	2.4	0.0029
	0.15	1391	5.8	0.0034
	0.25	1371	2.5	0.0036
LAFO				
	0.05	446	1.21	0.0012
	0.15	213	0.77	0.0015
	0.25	771	2.97	0.0010
LSFO				
	0.05	232415	21.5	0.2081
	0.10	5979	4.95	0.0015
	0.15	736	178	0.0006
	0.20	1911	955	0.0011

$b\sigma_{ac}$ - AC conductivity at 5 MHz.
LCFO - $La_{1-x}Ce_xFeO_3$
LZFO- $La_{1x}Zn_xFeO_3$
LAFO - $La_{1-x}Al_xFeO_3$
LSFO - $La_{1-x}Sr_xFeO_3$
Conc. - concentration

3.10 Ferroelectric properties of lanthanum orthoferrite-type multiferroics

The ferroelectric properties of the synthesized lanthanum orthoferrite-type multiferroics have been analyzed by the P-E hysteresis curves recorded at room temperature using a ferroelectric loop tracer.

3.10.1 Ce-substituted lanthanum orthoferrites - $La_{1-x}Ce_xFeO_3$

To investigate the effect of Ce-substitution in the ferroelectric behavior of $LaFeO_3$, the room temperature ferroelectric hysteresis loops were recorded at different applied electric fields. During ferroelectric measurements, the ferroelectric loop tracer could trace the P-E hysteresis loop for the pristine sample at a low applied electric field of 1 kV/cm but it could not trace the ferroelectric response for the Ce-substituted samples at this field, which may be due to the high resistance (or low ac conductivity) of the substituted samples. For these

Ce-substituted samples, the P-E loops were recorded at various high electric fields such as 15 kV/cm, 17 kV/cm, 20 kV/cm and 25 kV/cm.

Figures 2.61 (a)-(e) illustrate the P-E loops of $La_{1-x}Ce_xFeO_3$ (x=0.00, 0.03, 0.06, 0.09 & 0.12) multiferroics. For the pristine sample, the P-E loop was traced at 1 kV/cm field, 50 Hz frequency and for the Ce-substituted samples, they were traced at 20 kV/cm field, 20Hz frequency. The observed P-E hysteresis loops confirm the presence of ferroelectric ordering in all the samples. The ferroelectric parameters such as maximum electric polarization (P_m), remanent polarization (P_r) and electric coercive field (E_C) extracted from the hysteresis loops are given in Table 2.29. From Table 2.29, it is evident that for the Ce-substituted samples, the values of maximum electric polarization vary alternatively, which is contrary to the variation of maximum magnetization. This may be due to the antisymmetric exchange interaction between the magnetic (Fe^{3+}) ions and rare earth (La^{3+}) ions.

In $LaFeO_3$, the electric polarization is due to the space charge polarization, which arises from the rotational displacement of $Fe^{2+} \leftrightarrow Fe^{3+}$ dipoles (n-type charge carriers) and the local displacement of $La^{3+} \leftrightarrow La^{4+}$ ions (p-type charge carriers) [Thirumalairajan et al., 2015]. The rotational displacement of the $Fe^{2+} \leftrightarrow Fe^{3+}$ dipoles may be visualized as the electronic exchange (n-type charge carriers) between the $Fe^{2+} \leftrightarrow Fe^{3+}$ ions. Similarly, the local displacement of the $La^{3+} \leftrightarrow La^{4+}$ dipoles may be visualized as the hole exchange (p-type charge carriers) between the $La^{3+} \leftrightarrow La^{4+}$ ions. Hence, the rotationaldisplacements of n-type charge carriers and the local displacements of p-type charge carriers in the direction of the applied electric field results in the net electric polarization in $LaFeO_3$. Even though, p-type carriers contribute to the net electric polarization their contribution is smaller than the n-type carriers because the mobility of the former is smaller than the latter. For the sample with Ce concentration x=0.12, the values of maximum electric polarization (P_m) and remanent polarization (P_r) are found to be higher (P_m=1.09 $\mu C/cm^2$ and P_r=0.83 $\mu C/cm^2$) than, those values observed for x=0.09, which is due to the presence of more number of p-type charge carriers.

3.10.1.1 Correlation between charge density and ferroelectric properties of $La_{1-x}Ce_xFeO_3$ multiferroics

The alternative variation of bond critical point charge density values along the bonds Fe-O1 and La-O1 of the Ce-substituted samples is good evidence for the change in the values of both types of charge carriers, which are attributed to the variation of maximum electric polarization. Similarly, the relatively high value of BCP charge density along the La-O1 bond of the sample with Ce concentration x=0.12, is good evidence for the presence of more p-type charge carriers in this particular sample which is responsible for its high maximum electric polarization (P_m) and remanent polarization.

3.10.2 Zn-substituted lanthanum orthoferrites - $La_{1-x}Zn_xFeO_3$

To investigate the effect of Zn-substitution in the ferroelectric behavior of $LaFeO_3$, the polarization versus electric field (P-E) hysteresis loops were recorded at room temperature for different frequencies in the applied electric field. The ferroelectric loop tracer could trace the P-E hysteresis loops for all the prepared samples in the low electric field of 1 kV/cm and at the frequencies of 20 Hz and 50 Hz. Figures 2.62 (a)-(d) show P-E hysteresis loops of $La_{1-x}Zn_xFeO_3$ (x=0.00, 0.05, 0.15 & 0.25) multiferroics traced in the applied field of 1 kV/cm and at a frequency of 20 Hz. The traced P-E hysteresis loops illustrate the rectangular–like loops with high P_r, which reflect the electrical leakage in these materials as observed in $BiFeO_3$ [Basith et al., 2017]. Usually, the accumulation of stoichiometry defects during crystal growth, oxygen vacancies, domains with different orientations, and grain boundaries leads to electrical leakage in ceramic materials [Simõeset al., 2009]. Here, in the synthesized $La_{1-x}Zn_xFeO_3$ samples, the observed electrical leakage is mainly attributed to oxygen vacancies due to charge imbalance in the host lattice as discussed in XRD data analysis. The P-E loops also show the jerky change in polarization–the Barkhausen jumps, which arises due to rapid nucleation and orientation of domains [Rudyak, 1971, Chynoweth, 1958] in the regions that are stressed due to crystal inhomogeneities.

The ferroelectric parameters such as maximum electric polarization (P_m), remanent polarization (P_r) and electric coercive field (E_C) extracted from the P-E loopsare listed in Table 2.30. From Table 2.30, it is observed that the value of maximum electric polarization is found to increase slightly with the increase in Zn substitution. The electric polarization in $LaFeO_3$, is due to the rotational displacement of $Fe^{2+} \leftrightarrow Fe^{3+}$ dipoles (n-type charge carriers) and the local displacement of $La^{3+} \leftrightarrow La^{4+}$ ions (p-type charge carriers) [Thirumalairajan et al., 2015]. The appreciable values of bond critical point charge density along the bonds Fe-O1and La-O1 are good evidence for the presence of n-type and p-type charge carriers respectively, which may be attributed to the value of maximum electric polarization. Among the Zn-substituted samples, for x=0.25, the values of maximum electric polarization (P_m) and remanent polarization (P_r) are found to be relatively high (P_m =20.12 $\mu C/cm^2$ and P_r =20.12 $\mu C/cm^2$), which may be due to the presence of more n-type and p-type charge carriers.

3.10.2.1 Correlation between charge density and ferroelectric properties of $La_{1-x}Zn_xFeO_3$ multiferroics

The relatively high values of BCP charge density along the Fe-O1 and La-O1 bonds of the sample with x=0.25 are good evidence for the presence of more n-type and p-type charge carriers respectively, which are responsible for its high polarization.

3.10.3 Al-substituted lanthanum orthoferrites - $La_{1-x}Al_xFeO_3$

$LaFeO_3$ is an antiferromagnetic insulator, wherein ferroelectricity is induced by magnetism. The ferroelectricity in the polycrystalline ceramic $LaFeO_3$ is difficult to demonstrate due to the random orientation of crystallites. The P-E hysteresis loop of $LaFeO_3$ will be a rounded one [Vandeven et al., 1967] instead of square due to the orientation of crystallites. To analyze the effect of Al substitution on the ferroelectric properties of $La_{1-x}Al_xFeO_3$, the ferroelectric hysteresis (P-E) loops were traced at room temperature at an electric field of 1 kV/cm and different frequencies 20 Hz, 50 Hz and 100 Hz using Ferroelectric loop tracer. Figures 2.63 (a)-(c) show the P-E hysteresis loops of $La_{1-x}Al_xFeO_3$ (x=0.05, 0.15 & 0.25) multiferroics traced at an applied electric field of 1 kV/cm and a frequency of 20 Hz. The ferroelectric parameters such as maximum electric polarization (P_m), remanent polarization (P_r) and electric coercive field (E_C) obtained from the ferroelectric hysteresis loops are summarized in Table 2.31. Figures 2.63 (a)-(c) demonstrate elliptical P-E loops with high P_r, which reflects the electrical leakage in the synthesized samples [Basith et al., 2017]. It is known that the accumulation of stoichiometry defects during crystal growth, oxygen vacancies, domains with different orientations, and grain boundaries [Simões et al., 2009] can instigate severe electrical leakage in ceramics. Here, the electrical leakage in the synthesized $La_{1-x}Al_xFeO_3$ multiferroics is attributed to domains with different orientations, as discussed in the magnetic properties. From Table 2.31, it is observed that for the prepared $La_{1-x}Al_xFeO_3$ multiferroics, the values of P_m are high which reflects that the contribution of both types of charge carriers to the net polarization is significant. From Table 2.31, it is also seen that the maximum electric polarization (P_m) and remanent polarization (P_r) lie in the range of 96.53 to 151.4 $\mu C/cm^2$ and 96.5 to 151.3 $\mu C/cm^2$ respectively. It is well known that in $LaFeO_3$, the rotational displacement of $Fe^{2+} \leftrightarrow Fe^{3+}$ dipoles (*n*-type charge carriers) and the local displacement of $La^{3+} \leftrightarrow La^{4+}$ dipoles (p-type charge carriers) [Thirumalairajan et al., 2015] are responsible for its electric polarization. Among the synthesized $La_{1-x}Al_xFeO_3$ multiferroics, for *x*=0.15, the values of P_m and P_r are found to be relatively high as 151.38 and 151.26 $\mu C/cm^2$. The reason for the high values of P_m and P_r for x=0.15 is due to the contribution of p-type charge carriers to the net electric polarization is more in this substitution level.

3.10.3.1 Correlation between charge density and ferroelectric properties of La$_{1-x}$Al$_x$FeO$_3$ multiferroics

The high value of BCP charge density along the La-O1 bond of the system with Al concentration x=0.15 is good evidence for the presence of more p-type charge carriers that are responsible for its high ferroelectric property.

The ferroelectric parameters of the synthesized lanthanum orthoferrite-type multiferroics extracted from the ferroelectric hysteresis curves have been tabulated in Table 3.10.

Table 3.10 Ferroelectric parameters of lanthanum orthoferrite-type multiferroics

Samples	Conc. (x)	Applied electric field (kV/cm)	Frequency (Hz)	P$_m$ (µC/cm^2)	P$_r$ (µC/cm^2)	E$_C$ (kV/cm)
	0.00	1	50	6.83	3.42	0.7
	0.03	20	20	0.81	0.45	12
	0.06	20	20	0.86	0.59	14.3
LCFO	0.09	20	20	0.65	0.44	16.3
	0.12	20	20	1.09	0.83	15.3
	0.00	1	20	20.01	20.01	0.44
LZFO	0.05	1	20	20	19.26	0.94
	0.15	1	20	20.11	20.11	0.23
	0.25	1	20	20.12	20.12	0.21
LAFO	0.05	1	20	96.53	96.50	0.97
	0.15	1	20	151.4	151.3	0.81
	0.25	1	20	122	120	0.65

P$_m$ - Maximum electric polarization, P$_r$ - Remanent polarization, E$_C$ - Electric coercivefield
LCFO - La$_{1-x}$Ce$_x$FeO$_3$
LZFO - La$_{1-x}$Zn$_x$FeO$_3$
LAFO - La$_{1-x}$Al$_x$FeO$_3$
LSFO - La$_{1-x}$Sr$_x$FeO$_3$
Conc. - concentration

References

[1] Ahmed M.A., Azab A.A., El–Khawas E.H., J. Mater. Sci: Mater. Electron., 26, 8765 (2015).

https://doi.org/10.1007/s10854-015-3556-4

[2] Aroyo M.I., International Tables for crystallography Volume A: space- group symmetry, 2nd online edition, (2016). https://doi.org/10.1107/97809553602060000114

[3] Azab A.A., Helmy N., Albaaj S., Mater. Res. Bull., 66, 249 (2015). https://doi.org/10.1016/j.materresbull.2015.02.038

[4] Basith M.A., Billah A., Jalil M.A., Yesmin N., Sakib M.A., Ashik E.K., J. Alloys Compd., 694, 792 (2017). https://doi.org/10.1016/j.jallcom.2016.10.018

[5] Berkowitz A.E., Kentaro T., J. Magn. Magn. Mater., 200, 552 (1999). https://doi.org/10.1016/S0304-8853(99)00453-9

[6] Bhushan B., Wang Z., Tol J., Dalal N.S., Basumallick A., Nagasampagi Y., Kumar S., Das D., J. Am. Ceram. Soc., 95, 1985 (2012). https://doi.org/10.1111/j.1551-2916.2012.05132.x

[7] Chynoweth A.G., Phys. Rev., 110, 1316 (1958). https://doi.org/10.1103/PhysRev.110.1316

[8] Collins D.M., Nature, 298, 49 (1982). https://doi.org/10.1038/298049a0

[9] Cullity B.D., Stock S.R., Elements of X-ray diffraction, Pearson education. 3rd edition. Prentice Hall, Upper Saddle River, 558 (2001).

[10] Dantas A.L., Reboucas G.O.G., Silva A.S.W.T., Carrico A.S., J. Appl. Phys., 97, 10K105/ (1-3) (2005). https://doi.org/10.1063/1.1847931

[11] García Calderón R., Gómez Sal J.C., Iglesias J.R. , J. Magn. Magn. Mater., 701, 272-276 (2004). https://doi.org/10.1016/j.jmmm.2003.11.255

[12] Giannakas A.E., Leontiou A.A., Ladavos A.K., Pomonis P.J., Appl. Catal. A., 309, 254 (2006). https://doi.org/10.1016/j.apcata.2006.05.016

[13] Glazyrin K., McCammon C., Dubrovinsky L, Merlini M., Schollenbruch K., Woodland A., Hanfland M., Am. Mineral., 97, 128 (2012). https://doi.org/10.2138/am.2011.3862

[14] Goldschmidt V.M., Die Gesetze der Krystallochemie, DieNaturwissenschaften, 14, 477 (1926). https://doi.org/10.1007/BF01507527

[15] Gorelov B.M., Kotenok E.V., Makhno S.N., Sydorchuk V.V., Khalameida S.V., Zazhigalov V.A., Solid-State Electron., 56, 83 (2011). https://doi.org/10.1134/S1063784211010117

[16] Hearne G.R., Pasternak M.P., Phys. Rev. B: Condens. Matter., 51, 11495 (1995). https://doi.org/10.1103/PhysRevB.51.11495

[17] Holland T.J.B., Redfern S.A.T., Mineral Mag., 61, 65 (1997).
https://doi.org/10.1180/minmag.1997.061.404.07

[18] Ibers J.A., Hamilton W.C., International tables for X-ray crystallography, Vol. 4, Kynoch Press, Birmingham, (1974).

[19] Idrees M., Nadeem M., Atif M., Siddique M., Mehmood M., Hassan M.M., Acta Mater., 59, 1338 (2011). https://doi.org/10.1016/j.actamat.2010.10.066

[20] Izumi F., Dilanien R. A,. Recent Research Developments in Physics Part II, Vol.3, Transworld Research Network. Trivandrum, 699-726, (2002).

[21] JCPDS PDF# 37-1493

[22] JCPDS PDF# 11-0614

[23] JCPDS PDF# 43-1002

[24] Kodama R.H., Makhlouf S.A., Berkowitz A.E., Phys. Rev. Lett., 79, 1393 (1997).
https://doi.org/10.1103/PhysRevLett.79.1393

[25] Köferstein R., Jäger L., Ebbinghaus S.G., Solid State Ion, 249, 1 (2013).
https://doi.org/10.1016/j.ssi.2013.07.001

[26] Koops C.G., Phys. Rev., 83, 121(1951). https://doi.org/10.1103/PhysRev.83.121

[27] Li K., Wang D., Wu F., Xie T., Li T., Mater. Chem. Phys., 64, 269 (2000).
https://doi.org/10.1016/S0254-0584(99)00265-5

[28] Manzoor S., Husain S., J. Appl. Phys., 124, 065110-1 (2018).
https://doi.org/10.1063/1.5025252

[29] Maxwell J.C., Electricity and Magnetism, Oxford University Press, London (1973).

[30] Momma K., Izumi F., VESTA: a three-dimensional visualization system for electronic and structural analysis. J. Appl. Crystallogr. 41, 653 (2008).
https://doi.org/10.1107/S0021889808012016

[31] Nguyen T., Knurova M.V., Nguyen T.M., Mittova V.O., Mittova I.Ya,. Nanosystems: Phys. Chem. Math., 5, 692 (2014).

[32] Petricek V., Dusek M., Palatinus L., Kristallogr Z, Crystallographic Computing System JANA 2006: General features., 229, 345 (2014).
https://doi.org/10.1515/zkri-2014-1737

[33] Qi X.W., Zhou J., Yue Z.X., Gui Z., Li L.T., Ceram. Int., 29, 347 (2003).
https://doi.org/10.1016/S0272-8842(02)00119-0

[34] Richens D.T., The Chemistry of Aqua Ions, Wiley, 604 (1997)

[35] Rietveld H. M., J. Appl. Crystallogr., 2, 65 (1969).
https://doi.org/10.1107/S0021889869006558

[36] Rudyak V.M., Sov. Phys. Uspekhi., 13, 461 (1971).
https://doi.org/10.1070/PU1971v013n04ABEH004681

[37] Saravanan R., Mangaiyarkkarasi J., J.Mater.Sci.:Mater.Electron., 27, 2523 (2016).
https://doi.org/10.1007/s10854-015-4053-5

[38] Saravanan R., Grain software (Personal communication) (2008), http://phymat.in/.

[39] Saravanan R., Thenmozhi N., Yen-Pei Fu, Physica B: Physica of Condens. Matter.,
493, 25 (2016).

[40] Sasikumar S., Saravanan R., Saravanakumar S., J.Mater.Sci.:Mater.Electron., 29,
1198 (2018).

[41] Shannon R.D., Acta Crystallogr., 32, 751 (1976).
https://doi.org/10.1107/S0567739476001551

[42] Shen H., Cheng G., Wu A., Xu J., Zhao J., Phys. Status Solidi A., 206, 1420 (2009).
https://doi.org/10.1002/pssa.200824266

[43] Shikha P., Kang T.S., Randhawa B.S., J. Alloys Compd., 625, 336 (2015).
https://doi.org/10.1016/j.jallcom.2014.11.074

[44] Simões A.Z., Garcia F.G., Riccardi C.D.S., Mater. Chem. Phys., 116, 305 (2009).
https://doi.org/10.1016/j.matchemphys.2009.04.036

[45] Thirumalairajan S., Girija K., Mastelaro V.R., Ponpandian N., J. Mater. Sci.- Mater.
Electron., 26, 8652 (2015). https://doi.org/10.1007/s10854-015-3540-z

[46] Vandeven D., Galy J., Pouchard M., Hagenmuller P., Mat. Res. Bull., 2, 809 (1967).
https://doi.org/10.1016/0025-5408(67)90008-6

[47] Wagner K.W., Ann. Phys., 40, 818 (1993).

[48] Winkler E., Zyster R.D., Mansilla M.V., Fiorant D., Phys. Rev. B, 72, 132409-1-4
(2005). https://doi.org/10.1103/PhysRevB.72.132409

[49] Wood D. L., Tauc J., Phys Rev B. 5, 3144 (1972).
https://doi.org/10.1103/PhysRevB.5.3144

[50] Wyckoff R.W.G., Crystal structures, Vol.1, Inter-space publishers, London, (1963).

Chapter 4

Conclusion

In this work, the following four series of lanthanum orthoferrite (LFO)-type multiferroics have been synthesized by solid state reaction method.

(i) $La_{1-x}Ce_xFeO_3$ (x=0.00, 0.03, 0.06, 0.09 and 0.12) (LCFO)

(ii) $La_{1-x}Zn_xFeO_3$ (x=0.00, 0.05, 0.15, and 0.25) (LZFO)

(iii) $La_{1-x}Al_xFeO_3$ (x=0.05, 0.15, 0.25 and 0.35) (LAFO) and

(iv) $La_{1-x}Sr_xFeO_3$ (x=0.05, 0.10, 0.15 and 0.20) (LSFO)

The synthesized samples have been characterized using powder X-ray diffraction (PXRD), scanning electron microscopy (SEM), energy dispersive X-ray spectroscopy (EDS), UV-visible absorption spectroscopy (UV-Vis), vibrating sample magnetometry (VSM), dielectric measurements (Impedance analysis), ferroelectric measurements (polarization versus electric field (P-E) hysteresis loop). For all the synthesized multiferroics, the charge density distribution and bonding nature of atoms in the unit cell have been analyzed accurately using Maximum Entropy Method (MEM). The optical, magnetic, dielectric and ferroelectric properties have been analyzed using MEM-based BCP charge density values. The results obtained from the above experimental characterizations and analytical techniques are analyzed and summarized in this section.

(i) Structural analysis

The crystallographic phase formation and structural analysis of all the synthesized multiferroics have been done by experimental powder XRD patterns and XRD profile refinement study. The experimental XRD patterns and the profile refinement confirm that all the synthesized multiferroics have been crystallized in orthorhombic structure with space group *Pnma*.

The experimental XRD patterns of $La_{1-x}Ce_xFeO_3$ (x=0.00, 0.03, 0.06, 0.09 & 0.12) multiferroics show that the samples are monophasic except for x=0.09 and x=0.12, which have extra reflections corresponding to CeO_2. The refined cell parameters from unit cell refinement show that up to x=0.09, the orthorhombic unit cell parameters and the unit cell volume decrease with the increase in Ce concentration. The average grain size of the synthesized $La_{1-x}Ce_xFeO_3$ (x=0.00, 0.03, 0.06, 0.09 and 0.12) ranges from 57.9(2.4) nm to 88.1(1.3) nm.

The raw XRD patterns of $La_{1-x}Zn_xFeO_3$ (x=0.00, 0.05, 0.15 & 0.25) multiferroics show that the pristine sample is monophasic and the samples with Zn concentration x=0.05, 0.15, and 0.25 have additional peak corresponding to Fe_3O_4. The refined cell parameters from unit cell refinement indicate that the unit cell parameters and also the unit cell volume of the samples non-monotonically decrease with the increase in Zn concentration. For the prepared $La_{1-x}Zn_xFeO_3$ (x=0.00, 0.05, 0.15, and 0.25) multiferroics, the average grain size non-monotonically decreases with the increase in Zn concentration and it ranges from 25.9(0.5) nm to 36.7(1) nm.

The raw XRD patterns of $La_{1-x}Al_xFeO_3$ (x=0.05, 0.15 & 0.25) multiferroics show that the XRD peaks shift towards the higher 2θ angles with the increase in Al concentration in the host lattice. The refined structural and profile fitted parameters extracted from the Rietveld refinement method show that the cell parameters and unit cell volume decrease with the increase in the substitution of Al. The average grain size of the synthesized $La_{1-x}Al_xFeO_3$ (x=0.05, 0.15 & 0.25) multiferroics ranges from 26.1(0.6) nm to 37.7(1.7) nm.

The powder XRD patterns of $La_{1-x}Sr_xFeO_3$ (x=0.05, 0.10, 0.15 & 0.20) multiferroics confirm that all the samples are monophasic exhibiting orthogonal symmetry with space group *Pnma*. The structural and profile fitted parameters obtained from the Rietveld refinement method show that the unit cell parameters and cell volume decrease with the increase in Sr concentration. The average grain size of the synthesized $La_{1-x}Sr_xFeO_3$ (x=0.05, 0.10, 0.15 & 0.20) multiferroics ranges from 21.4(1.2) nm to 26.1(0.8) nm.

(ii) Surface morphology, microstructure and elemental composition analysis

The surface morphology and microstructure of the synthesized multiferroics have been analyzed using scanning electron microscope (SEM) images. The SEM micrographs of the synthesized multiferroics reveal that the formed crystals have numerous deagglomerated particles that are heterogeneously distributed with irregular shapes and different sizes. The average particle size of the synthesized multiferroics is found to be in the range from 0.64(0.2) μm to 1.34(0.6) μm. The micrometer-sized particles confirm that all the synthesized samples are bulk multiferroics.

The purity and the elemental compositions of the synthesized multiferroics have been analyzed by EDS analysis. EDS spectra of all the synthesized samples depict various peaks belonging to the constituent elements of the samples. The numerical values of atomic and weight percentages of the various elements present in the synthesized multiferroics reveal that all the samples contain only the constituent atomic species without impurities.

(iii) Charge density analysis

The charge density distribution and primary chemical bonding between the constituent atoms in the unit cell of all the synthesized multiferroics have been systematically analyzed by the MEM-based charge density study. The qualitative and quantitative analysis of charge density reveals that the bond Fe-O1 is predominantly covalent in nature and the bond La-O1 is equally covalent in nature for the synthesized multiferroics.

The charge density study for $La_{1-x}Ce_xFeO_3$ reveals that for the sample with Ce concentration x=0.06, the values of bond critical point charge density along the Fe-O1 bond and La-O1 bond are found to be relatively high as 0.834 e/$Å^3$ and 1.0741 e/$Å^3$ when compared with other concentrations, which confirms that the bond Fe-O1 is more covalent and the bond La-O1 is also more covalent with slightly more charge density in this substitution level.

The charge density study for $La_{1-x}Zn_xFeO_3$ reveals that for the Zn-substituted samples, the BCP charge density values along the primary bonds Fe-O1 and La-O1 are found to be less than those for x=0.00. Among the Zn-substituted samples, for the sample with Zn concentration x=0.25, the BCP charge density values along the primary bonds Fe-O1 and La-O1 are found to be high as 0.5929 e/$Å^3$ and 0.5671 e/$Å^3$, which confirms that the bond Fe-O1 is more covalent and the bond La-O1 is also more covalent with slightly more charge density in this substitution level.

The charge density study for $La_{1-x}Al_xFeO_3$ reveals that for x=0.25, the BCP chargedensity along the Fe-O1 bond is found to be relatively high as 1.015 e/$Å^3$, which confirmsthat the bond Fe-O1 is more covalent in this substitution level. For the synthesized $La_{1-x}Al_xFeO_3$, it is found that for the La-O1 bond, the BCP charge densities vary from 0.3175 e/$Å^3$ to 0.4699 e/$Å^3$. The low values of BCP charge density declare that the bond La-O1 is less covalent in character. Among the prepared samples, for x=0.15, the BCP charge density is found to be relatively high as 0.4699 e/$Å^3$, which confirms that the bond La-O1 is more covalent with slightly more charge density in this substitution level.

The charge density analysis for $La_{1-x}Sr_xFeO_3$ reveals that the BCP charge densities along the bond Fe-O1 are found to be high, and they are in the range from 0.7782 e/$Å^3$ to 0.8693 e/$Å^3$, which declares that the bond Fe-O1 is predominantly covalent in character. From the charge density analysis, it is seen that for the La-O1 bond, the charge density values are high due to the spatial distribution of charges from neighboring atoms, and they are found to vary from 0.5281 e/$Å^3$ to 1.0203 e/$Å^3$, which declares that the bond La-O1 is equally covalent. Among the synthesized $La_{1-x}Sr_xFeO_3$ (x=0.05, 0.10, 0.15 & 0.25), for the sample with x=0.05, the values of BCP charge density along the bonds Fe-O1 and La-O1 are found to be relatively high as 0.8693 e/$Å^3$ and 1.0203 e/$Å^3$, which confirm that the bond

Fe-O1 is more covalent and the bond La-O1 is also more covalent with slightly more charge density in this substitution level.

(iv) Optical properties

The optical properties of all the synthesized multiferroics have been analyzed using UV-visible absorption spectra. Using the Tauc plot, the energy band gap has been estimated. The energy band gap increases with the increase in the substitution of Ce, Zn and Sr, whereas it decreases with the increase in the substitution of Al. The decrease in the BCP charge density values along the Fe-O1 bond of Ce, Zn and Sr-substituted samples is attributed to the decrease in the number of Fe^{3+} ions with uncompensated spins, which are responsible for the increase in the energy band gap values. Also, the increase in the BCP charge density values along the bond Fe-O1 with the increase in the substitution of Al is attributed to the increase in the number of uncompensated spins of Fe^{3+} ions, which are responsible for the decrease in the energy band gap values.

(v) Magnetic properties

The magnetic properties of all the synthesized multiferroics have been analyzed using the magnetic hysteresis (M-H) curves recorded at room temperature by the vibrating sample magnetometer (VSM).

The M-H curves of $La_{1-x}Ce_xFeO_3$ (x=0.00, 0.03, 0.06, 0.09 & 0.12) multiferroics show that all the samples exhibit weak ferromagnetic (FM) behavior with appreciable magnetization due to the spin canting of antiferromagnetically ordered Fe moments through Fe^{3+}–O^{2-}–Fe^{3+} super-exchange interaction. The negative exchange bias is observed in $La_{1-x}Ce_xFeO_3$ (x=0.00, 0.03, 0.06, 0.09 & 0.12) multiferroics. The magnetic properties of Ce-substituted samples are found to be enhanced due to the presence of uncompensated Fe^{3+} spins caused by the decrease in particle size, which leads to the strengthened antisymmetric exchange interaction. The significant values of BCP charge density along the bonds Fe-O1 and La-O1 for the Ce-substituted samples are good evidence for the presence of Fe^{3+} spins and strengthened antisymmetric exchange interaction. For the sample with Ce concentration x=0.06, the values of H_C and H_{EB} are found to be relatively high when compared with other Ce-substituted samples, which is due to the probability of formation of AFM domains is more than that of FM domains in this level of substitution. For x=0.09, saturation magnetization (M_s) and remanent magnetization (M_r) are found to be higher than those observed for x=0.12, which is due to the presence of more Fe^{3+} spins. The relatively high value of charge density along the Fe-O1 bond of the sample with x=0.09 is good evidence for the presence of more Fe^{3+} spins which are responsible for its high saturation and remanent magnetizations.

The M-H curves of the synthesized $La_{1-x}Zn_xFeO_3$ (x=0.00, 0.05, 0.15 & 0.25) multiferroics show narrow-width hysteresis curves, which reflect that the prepared samples are soft in nature with weak ferromagnetic (FM) behavior. The ferromagnetic ordering in $La_{1-x}Zn_xFeO_3$ is promoted by the uncompensated spins of Fe^{3+} ions which causes the spin canting of Fe moments by means of super-exchange interaction between Fe metal atoms via oxygen atoms. The M-H curves also show negative exchange bias (EB) in the synthesized samples. The magnetic parameters M_r, H_C, and H_{EB} are found to decrease, whereas the saturation magnetization (M_s) is found to increase for the Zn-substituted samples when compared with pure $LaFeO_3$, which is due to the probability of formation of FM domains is more than that of AFM domains. For x=0.25, the value of saturation magnetization is found to be relatively high, which is due to the presence of more uncompensated spins of Fe^{3+} ions which strengthens antisymmetric exchange interaction. Furthermore, for x=0.25, the value of H_{EB} is found to be low which is attributed due to the decrease of magnetocrystalline anisotropy that arises as a result of the increase in the FM spins. The relatively high values of BCP charge density along the bonds Fe-O1 and La-O1 for x=0.25 are good evidence for the presence of more uncompensated Fe^{3+} spins and strengthened antisymmetric exchange interaction respectively, which are responsible for its enhanced magnetic behavior.

The M-H curves of the synthesized $La_{1-x}Al_xFeO_3$ (x=0.05, 0.15 & 0.25) multiferroics reveal that all the samples behave like a weak ferromagnet. The weak ferromagnetism is originated from canting of antiferromagnetically ordered metal ion (Fe^{3+}) spins that arise as a result of $Fe^{3+}-O^{2-}-Fe^{3+}$ super-exchange interaction. The magnetic hysteresis curves also illustrate the jump-like magnetization at zero magnetic fields which are termed Barkhausen jumps. The magnetic parameters increase with the increase in the substitution of Al due to the increase of uncompensated spins of Fe^{3+} ion which results in significant antisymmetric exchange interaction with the La^{3+} ions. The increase in the values of BCP charge density along the bond Fe-O1 with the increase in Al concentration is good evidence for the increase in the number of uncompensated spins of Fe^{3+} ions and the significant values of BCP charge density along the bond La-O1 are good evidence for the significant antisymmetric exchange interaction between the Fe^{3+} and La^{3+} ions which are responsible for the magnetic behavior of $La_{1-x}Al_xFeO_3$ multiferroics.

The M-H curves of the synthesized $La_{1-x}Sr_xFeO_3$ (x=0.05, 0.10, 0.15 & 0.20) multiferroics show that all the samples exhibit ferromagnetic (FM) behavior with appreciable magnetization and coercivity. The weak ferromagnetism can be explained by the canting of antiferromagnetically ordered metal ion (Fe^{3+}) spins. For the sample with Sr concentration x=0.05, the values of magnetic parameters such as M_s, M_r and H_C are found to be relatively high, which is due to the presence of more Fe^{3+} ions with uncompensated spins which strengthens the antisymmetric exchange interaction between the Fe^{3+} ions and the La^{3+} ions. The high values of BCP charge density along the bonds Fe-O1 and La-O1 for x=0.05 are good evidence for the presence of more Fe^{3+} ions with uncompensated spins and strengthened antisymmetric exchange interaction respectively, which are responsible for its enhanced magnetic behavior.

(vi) Dielectric properties

The dielectric measurements of all the synthesized multiferroics have been carried out in the wide frequency range from 10 Hz to 5 MHz at room temperature using an impedance analyzer. All the synthesized multiferroics exhibit the normal dielectric dispersion which is attributed to space charge polarization.

The dielectric measurements of $La_{1-x}Ce_xFeO_3$ show that for pristine $LaFeO_3$, the dielectric constant (ε') and ac conductivity are 2.7 x 10^4 and 0.3389 $\Omega^{-1}m^{-1}$, which are higher than those for Ce-substituted samples. The dielectric loss (tan δ) for the pristine sample is observed as 9.4 and for the Ce-substituted samples, it is found to vary from 3.5 to 41. The low dielectric property and ac conductivity in Ce-substituted samples is attributed to the decrease in the number of Fe^{3+} ions which are responsible for both space charge polarization and hopping of charge carriers between the localized states. Among the Ce-substituted samples for x=0.06, the values of ε' and ac conductivity are found tobe relatively high, which is due to the presence of more cations (Fe^{3+} and La^{3+} ions) with n-type and p-type charge carriers. The high values of BCP charge density along the bonds Fe-O_1 and La-O_1 for the sample with Ce concentration x= 0.06, reveal the presence of more charge carriers that are responsible for its high dielectric constant and ac conductivity.

The dielectric measurements of $La_{1-x}Zn_xFeO_3$ show that for pure $LaFeO_3$, the value of dielectric constant (ε') is about 9.55 x 10^2, with high dielectric loss (tan δ) of about 5.9 and its ac conductivity is relatively low as 0.0028 $\Omega^{-1}m^{-1}$. Among the Zn-substituted samples, for x=0.25, the values of dielectric constant and ac conductivity are found to be appreciably high with low dielectric loss. The high values of ε' and σ_{ac} forthis sample (x=0.25) are attributed to the presence of more uncompensated spins of Fe^{3+} ions which are responsible for both space charge polarization and hopping of charge carriers between the localized states. The high BCP charge density values along the bondsFe-O1 and La-O1 for x=0.25 are good evidence for the presence of more uncompensated spins of Fe^{3+} ions and hence charge carriers that are responsible for its high ε' and σ_{ac}.

The dielectric measurements of the prepared $La_{1-x}Al_xFeO_3$ show that the dielectric constant (ε') and dielectric loss (tan δ) lie in the range of 213 to 771 and 0.77 to 2.97 respectively. The ac conductivity measurements indicate that the ac conductivity (σ_{ac}) of $La_{1-x}Al_xFeO_3$ ranges from 1.07 x 10^{-3} $\Omega^{-1}m^{-1}$ to 1.51 x 10^{-3} $\Omega^{-1}m^{-1}$. For x=0.25, the value of dielectric constant is found to be high as 771, due to high space charge polarization caused by the rotational displacement of more $Fe^{2+} \leftrightarrow Fe^{3+}$ dipoles. The more $Fe^{2+} \leftrightarrow Fe^{3+}$ dipoles in this sample (x=0.25) are due to the increase in the number of uncompensated spins of Fe^{3+} ions. The high BCP charge density along the Fe-O1 bond for this system (x=0.25) is good evidence for the increase in the number of uncompensated spins of Fe^{3+} ions and hence more $Fe^{2+} \leftrightarrow Fe^{3+}$ dipoles. From the ac conductivity measurement, it is seen that for x=0.15, ac conductivity is slightly high as $1.51 \times 10^{-3} \, \Omega^{-1}m^{-1}$, which is attributed to the contribution of p-type charge carriers to the electrical conduction is more in this substitution level. The high BCP charge density along the La-O1 bond for this (x=0.15) system is good evidence for the presence of more p-type charge carriers, which are responsible for its high conductivity.

The dielectric measurements of the synthesized $La_{1-x}Sr_xFeO_3$ (x=0.05, 0.10, 0.15 & 0.20) multiferroics show that the dielectric property and ac conductivity decreases with the increase in Sr concentration. For x=0.05, the ε' and σ_{ac} are found to be relatively high as 2.32×10^5 and $0.2081 \, \Omega^{-1}m^{-1}$. The giant dielectric constant and high conductivity observed for this sample (x=0.05) is attributed to the presence of more cations (Fe^{3+} and La^{3+} ions) with n-type and p-type charge carriers. The high values of charge density along the bonds Fe-O1 and La-O1 for x=0.05, are good evidence for the presence of more charge carriers which results in its high dielectric constant and ac conductivity. For the synthesized $La_{1-x}Sr_xFeO_3$, the dielectric loss (tan δ) lies in the range of 5 to 955. Since the dielectric constant for x=0.05, is found to be relatively high, the dielectric loss factor corresponding to this concentration can be considered as low.

(vii) Ferroelectric properties

The ferroelectric properties of the synthesized lanthanum orthoferrite-type multiferroics have been analyzed by the P-E hysteresis curves traced at room temperature using the ferroelectric loop tracer.

The P-E loops of $La_{1-x}Ce_xFeO_3$ (x=0.00, 0.03, 0.06, 0.09 & 0.12) multiferroics confirm the presence of ferroelectric ordering in all the samples. For the Ce-substituted samples, the values of maximum electric polarization vary alternatively. The alternative variation of bond critical point charge density values along the bonds Fe-O1 and La-O1 of the Ce-substituted samples is good evidence for the change in the values of both types of charge carriers which may be attributed to the variation of maximum electric polarization. For the sample with Ce concentration x=0.12, the values of maximum electric polarization (P_m) and remanent polarization (P_r) are found to be higher than those observed for x=0.09, which is due to the presence of more p-type charge carriers. The relatively high value of

BCP charge density along the La-O1 bond of the sample with Ce concentration x=0.12 is good evidence for the presence of more p-type charge carriers in this particular sample.

The P-E hysteresis loops of $La_{1-x}Zn_xFeO_3$ (x=0.00, 0.05, 0.15 & 0.25) multiferroics illustrate the rectangular–like loops with high P_r which reflect the electrical leakage in these materials. The observed electrical leakage is mainly attributed to oxygen vacancies due to charge imbalance in the host lattice. The P-E loops also show the jerky change in polarization–called Barkhausen jumps, which arises due to rapid nucleation andorientation of domains in the stressed regions of crystal. The value of maximum electric polarization is found to increase slightly with the increase in Zn substitution. For x=0.25, the values of maximum electric polarization (P_m) and remanent polarization (P_r) are found to be relatively high, which is due to the presence of more n-type and p-type charge carriers. The relatively high values of BCP charge density along the Fe-O1 and La-O1 bonds for this sample (x=0.25) are good evidence for the presence of more n-type and p-type charge carriers respectively.

The P-E hysteresis loops of $La_{1-x}Al_xFeO_3$ (x=0.05, 0.15 & 0.25) multiferroics demonstrate the elliptical P-E loops with high P_r, which reflects the electrical leakage in the synthesized samples. Here, the observed electrical leakage has been attributed to domains with different orientations. For the prepared $La_{1-x}Al_xFeO_3$, the values of P_m are high which reflects that the contribution of both types of charge carriers (n-type and p-type) to the net polarization is significant. Among the synthesized samples, for Al concentration x=0.15, the values of P_m and P_r are found to be relatively high, which is dueto the contribution of p-type charge carriers is more to the net electric polarization in this substitution level. The high value of BCP charge density along the La-O1 bond of this system is good evidence for the presence of more p-type charge carriers and hence its high ferroelectric property.

To conclude, in this work, the charge density distribution and chemical bonding of lanthanum orthoferrite-type multiferroics have been analyzed in detail throughthe MEM-based charge density study. Also, the effects of various elements substitution on the optical, magnetic, dielectric and ferroelectric properties of lanthanum orthoferrite- type multiferroics have been analyzed through charge densities. Among the synthesized LFO-type multiferroics, $La_{1-x}Ce_xFeO_3$ and $La_{1-x}Zn_xFeO_3$ exhibit negative exchange bias with soft magnetic property and $La_{1-x}Al_xFeO_3$ and $La_{1-x}Sr_xFeO_3$ exhibit hard ferromagnetic property. Since the sample with Ce concentration x=0.06 exhibits large exchange bias and exchange anisotropy, it would be quite interesting in different switching and spin valve devices. Likewise, the sample with Zn concentration x=0.25 shows soft magnetic nature with high magnetization and low coercivity and also exhibits significant exchange bias, it can be used in magnetic storage media and novel spintronic device applications.

Among the synthesized $La_{1-x}Al_xFeO_3$ and $La_{1-x}Sr_xFeO_3$ multiferroics, $La_{1-x}Al_xFeO_3$ (x=0.25) exhibits higher saturation magnetization, remanent magnetization and coercivity than $La_{1-x}Sr_xFeO_3$ (x=0.05), which could make the system (x=0.25) a good candidate for magnetic memory device applications. Likewise, as the sample with Al concentration x=0.05, possesses a low dielectric constant and low dielectric loss at the high-frequency region, it can be recommended for use in high-frequency device applications such as information storage devices, sensors and spintronics. Furthermore, the multiferroic $La_{1-x}Sr_xFeO_3$ (x=0.05) is found to possess giant dielectric constant, high ac conductivity and low dielectric loss, it would be desirable for use in capacitors and energy storage devices in microelectronics.

Multiferroic Materials
Materials Research Foundations **140** (2023)

Materials Research Forum LLC
https://doi.org/10.21741/9781644902271

About the Author

Dr. **Ramachandran Saravanan** has been associated with the Department of Physics, The Madura College, affiliated with the Madurai Kamaraj University, Madurai, Tamil Nadu, India from the year 2000. He is the head of the Research Centre and PG department of Physics. He worked as a research associate during 1998 at the Institute of Materials Research, Tohoku University, Sendai, Japan and then as a visiting researcher at Centre for Interdisciplinary Research, Tohoku University, Sendai, Japan up to 2000.

Earlier, he was awarded the Senior Research Fellowship by CSIR, New Delhi, India, during Mar. 1991 - Feb.1993; awarded Research Associateship by CSIR, New Delhi, during 1994 – 1997. Then, he was awarded a Research Associateship again by CSIR, New Delhi, during 1997- 1998. Later he was awarded the Matsumae International Foundation Fellowship in1998 (Japan) for doing research at a Japanese Research Institute (not availed by him due to the simultaneous occurrence of other Japanese employment).

He has guided twelve Ph.D. scholars as of 2018, and about six researchers are working under his guidance on various research topics in materials science, crystallography and condensed matter physics. He has published around 150 research articles in reputed Journals, mostly International, apart from around 50 presentations in conferences, seminars and symposia. He has also guided around 60 M.Phil. scholars and an equal number of PG students for their projects. He has attracted government funding in India, in the form of Research Projects. He has completed two CSIR (Council of Scientific and Industrial Research, Govt. of India), one UGC (University Grants Commission, India) and one DRDO (Defense Research and Development Organization, India) research projects successfully and is proposing various projects to Government funding agencies like CSIR, UGC and DST.

He has written 12 books in the form of research monographs including; "Experimental Charge Density - Semiconductors, oxides and fluorides" (ISBN-13: 978-3-8383-8816-8; ISBN-10:3-8383-8816-X), "Experimental Charge Density - Dilute Magnetic Semiconducting (DMS) materials" (ISBN-13: 978-3-8383-9666-8; ISBN-10: 3-8383-9666-9) and "Metal and Alloy Bonding - An Experimental Analysis" (ISBN -13: 978-1-4471-2203-6). He has committed to write several books in the near future.

His expertise includes various experimental activities in crystal growth, materials science, crystallographic, condensed matter physics techniques and tools as in slow evaporation, gel, high temperature melt growth, Bridgman methods, CZ Growth, high vacuum sealing etc. He and his group are familiar with various equipment such as: different types of cameras; Laue, oscillation, powder, precession cameras; Manual 4-circle X-ray diffractometer, Rigaku 4-circle automatic single crystal diffractometer, AFC-5R and AFC-

7R automatic single crystal diffractometers, CAD-4 automatic single crystal diffractometer, crystal pulling instruments, and other crystallographic, material science related instruments. He and his group have sound computational capabilities on different types of computers such as: IBM – PC, Cyber180/830A – Mainframe, SX-4 Supercomputing system – Mainframe. He is familiar with various kind of software related to crystallography and materials science. He has written many computer software programs himself as well. Around twenty of his programs (both DOS and GUI versions) have been included in the SINCRIS software database of the International Union of Crystallography.

www.ingramcontent.com/pod-product-compliance
Lightning Source LLC
Chambersburg PA
CBHW071215210326
41597CB00016B/1828